Studies in Systems, Decision and Control

Volume 506

The series "Studies in Systems, Decision and Control" (SSDC) covers both new developments and advances, as well as the state of the art, in the various areas of broadly perceived systems, decision making and control–quickly, up to date and with a high quality. The intent is to cover the theory, applications, and perspectives on the state of the art and future developments relevant to systems, decision making, control, complex processes and related areas, as embedded in the fields of engineering, computer science, physics, economics, social and life sciences, as well as the paradigms and methodologies behind them. The series contains monographs, textbooks, lecture notes and edited volumes in systems, decision making and control spanning the areas of Cyber-Physical Systems, Autonomous Systems, Sensor Networks, Control Systems, Energy Systems, Automotive Systems, Biological Systems, Vehicular Networking and Connected Vehicles, Aerospace Systems, Automation, Manufacturing, Smart Grids, Nonlinear Systems, Power Systems, Robotics, Social Systems, Economic Systems and other. Of particular value to both the contributors and the readership are the short publication timeframe and the world-wide distribution and exposure which enable both a wide and rapid dissemination of research output.

Indexed by SCOPUS, DBLP, WTI Frankfurt eG, zbMATH, SCImago.

All books published in the series are submitted for consideration in Web of Science.

Jiayue Sun · Meina Zhai · Shu Liu · Peng Shi

Building a Low-Carbon Future

Adaptive Control Strategies for Distributed Energy Networks

Springer

Jiayue Sun
State Key Laboratory of Synthetical Automation for Process Industries and College of Information Science and Engineering
Northeastern University
Shenyang, Liaoning, China

Shu Liu
College of Information Science and Engineering
Northeastern University
Shenyang, Liaoning, China

Meina Zhai
College of Information Science and Engineering
Northeastern University
Shenyang, Liaoning, China

Peng Shi
School of Electrical and Mechanical Engineering
University of Adelaide
Adelaide, SA, Australia

ISSN 2198-4182 ISSN 2198-4190 (electronic)
Studies in Systems, Decision and Control
ISBN 978-3-031-43497-6 ISBN 978-3-031-43498-3 (eBook)
https://doi.org/10.1007/978-3-031-43498-3

This Springer imprint is published by the registered company Springer Nature Switzerland AG
The registered company address is: Gewerbestrasse 11, 6330 Cham, Switzerland

Paper in this product is recyclable.

To My Family
Jiayue Sun

To My Family
Meina Zhai

To My Family
Shu Liu

To My Family
Peng Shi

Preface

As countries around the world commit to achieving carbon neutrality by 2050 and pursue carbon peaking and carbon neutrality through the development of clean energy technologies, the power network is rapidly evolving toward a "distributed energy + Internet" direction, which has triggered a new wave of industrial revolution and energy reform and offered a new perspective for energy development. The distributed generator has garnered significant attention as a way to generate electricity with efficient use of renewable energy. Unlike traditional radial grid power sources arranged away from the load, distributed generators are commonly arranged around the load and can utilize decentralized energy sources such as wind, solar, and tidal energy, among others. They can also be small diesel or gas turbine power supply systems, but these are not preferred due to their carbon emissions. A microgrid is an effective grid organization that integrates various distributed power sources and provides power to local loads. It can effectively manage distributed power sources and local loads in an integrated manner, usually in grid-connected mode. However, it can also be planned or temporarily operated in isolation (or autonomy) on an emergency basis. With features such as tight regulation and start-up, multi-energy complementation, optimal management, and coordinated control, the microgrid is a necessary technical means for distributed power generation access and improving the reliability of power supply to customers while also reducing carbon emissions.

This monograph delves deeply into the output, load, and bus voltage regulation problems of distributed power supplies based on event-triggered and fault-tolerant control. Specifically, it investigates the following problems: (1) Cooperative event-triggered fault-tolerant voltage restoration in islanded AC microgrids; (2) fully distributed fault-tolerant event-triggered secondary control of microgrids under directed graphs; (3) output-feedback consensus fault-tolerant control with intermittent communication to achieve load voltage regulation; (4) secondary load voltage and frequency fault-tolerant regulation control via hierarchical mechanism; (5) distributed critical bus voltage regulation control for multi-microgrids with positive minimum inter-event times. Throughout the monograph, simulation results including some typical real applications are presented to illustrate the effectiveness and practicality of the cooperative control design methods proposed in previous sections in reducing carbon emissions and promoting sustainable energy development. The results also demonstrate the potential of microgrids to contribute to the global effort to reach carbon peaking and carbon neutrality by 2050.

Shenyang, China — Jiayue Sun
Shenyang, China — Meina Zhai
Shenyang, China — Shu Liu
Adelaide, Australia — Peng Shi
April 2023

Acknowledgements We would like to acknowledge (i) Profs. Qiuye Sun, Huaguang Zhang, Dr. Rui Wang, Bingyu Wang, Zhenwei Liu, and Dr. Jie Hu for collaborations on related work in this monograph;

The writing of this book was partially supported by the National Natural Science Foundation of China (62203097) and the China Postdoctoral Science Foundation (2022M710636).

Contents

List of Figures

Chapter 1
Introduction

As countries around the world commit to achieving carbon neutrality by 2050 and pursue carbon peaking and carbon neutrality through the development of clean energy technologies, microgrid (MG) technology has emerged as an essential component of the smart grid. Microgrid (MG) is an essential component of the smart grid. It integrates renewable energy sources, energy storage systems, and local loads to form a relatively small generation and distribution system suitable for a small range of low to medium voltages. It facilitates the effective integration of DGs [1–8]. It comprises various miniature power sources, loads, energy storage devices, and corresponding control and protection systems [9, 10].

The grid-connected MG operates in parallel with the distribution network through the power conversion device at the points of common coupling (PCC). When a fault occurs in the main grid, the MG can be cut off at the PCC and enter the islanding mode of operation. In addition, the coordinated power output between DG units and energy storage devices and the supporting control and protection links maintain the power balance within the MG to ensure a safe and reliable power supply for essential loads.

Although MGs provide conditions for large-scale application of DGs, the safe and stable operation within MGs still faces significant challenges, especially for independent MGs operating in island mode without main grids for power support, which need to rely on autonomous control of DGs within MGs to stabilize system voltage and frequency and maintain power balance. The DGs in MGs have various forms, such as photovoltaic cells, wind turbines, small gas turbines, fuel cells, and flywheel energy storage. Moreover, compared with traditional power grids powered by large-capacity generators, MG has fast response characteristics, low inertia and low overload capacity.

Rational control of the converter interfaces of DGs and coordination of the power output of different DGs are the keys to achieving a highly reliable load supply within the MG and ensuring the stability of MG operation. Thanks to computer technology and artificial intelligence development, distributed control based on multi-agent con-

J. Sun et al., *Building a Low-Carbon Future,* Studies in Systems, Decision and Control 506, https://doi.org/10.1007/978-3-031-43498-3_1

sensus algorithms provides a feasible solution for the reliable cooperative autonomy of MGs [11–16]. It is characterized by global information sharing using communication among DGs. Compared with traditional centralized and decentralized control, distributed control eliminates the problem of a single point of failure of the central controller. It distributes the communication and computation burden of the central controller to the local controllers of each DG, taking into account the advantages of the high reliability of decentralized control and global cooperative management of the centralized control to meet the plug in and plug out (PIPO) requirements of the MG.

1.1 Basic Structure of MG

According to the different power supply systems, MGs can be divided into three types: AC MG, DC MG and hybrid AC/DC MG.

AC MG is the most basic form of MG, which uses an AC bus as the busbar to connect various DG units, energy storage systems and load units. The DG, energy storage and DC loads in an AC MG must be connected to the grid through multi-stage AC/DC converters, resulting in low system efficiency. The advantage is that it does not need to change the original grid structure. It can be linked to the higher-level distribution network through PCC, which can be flexibly switched between grid-connected and isolated modes.

The DGs and loads in the DC MG are connected through a DC busbar, which can centrally utilize DC-type micropower and DC-type loads with less energy conversion process, high efficiency, low loss, no control problem on voltage phase and frequency, and high power quality and flexible reconfiguration. However, the construction of a DC MG needs to transform the original grid, and the protection technology of the DC system is less mature than the AC system.

AC-DC MG is a hybrid MG system containing AC and DC MGs. Each distributed power source, energy storage device, and load can be connected to their corresponding power supply system busbar. AC-DC MG can meet the power demand of various AC-DC loads with less energy conversion, featuring high operational efficiency, strong controllability and large grid connection capacity.

1.2 Hierarchical Control Structures

Recently, a hierarchical control structure, including primary, secondary, and tertiary control, has been proposed to ensure the global stability of MG [17–20].

1.2.1 Primary Control

The primary control layer is the local control layer, which maintains the system's essential supply and demand balance and ensures reliable operation. The most commonly used control method in primary control is droop control, in which communication-free current distribution can be achieved by simply setting the droop factor to maintain the voltage stability of the MG. Since the droop control does not depend on communication, the response time is fast, and the decentralized autonomy of DG is realized. Droop control meets the PIPO requirement of MGs and is widely used in the primary control of independent MGs. However, the droop control is a differential control where the system voltage and frequency are always lower than the reference values. Therefore, a secondary control needs to be introduced for regulation.

1.2.2 Secondary Control

The secondary control layer is the system regulation layer, which optimizes power quality and improves control accuracy. Droop control causes the system voltage to deviate from the reference value and requires secondary control to correct the output voltage. Any deviation in the output voltage of the DG caused by the primary control action is compensated. For AC MGs, secondary control aims to restore the system voltage to its reference value while achieving frequency restoration control and reactive power sharing.

Secondary control relies on communication between DGs to achieve co-regulation of state variables. Secondary control can be divided into centralized, distributed, and decentralized depending on the communication method. Decentralized control may not achieve reliable secondary control based only on the DG's information. Distributed control based on a multi-agent system (MAS) has the characteristics of high reliability and low cost. Distributed control can exchange information more efficiently than decentralized control [21–25]. Researchers have recently applied MAS theory to secondary control to compensate for the deficiencies of centralized and decentralized control and improve MG's stability and reliability. Secondary controller communication can correct the voltage fluctuations generated by droop control. This monograph focuses on the distributed secondary control strategy for islanded MG.

Centralized control uses the central controller to communicate with all DGs to collect global information and issue control commands after centralized processing [26]. The advantage of centralized control is that the central controller can be used to gather global information. Thus it can realize complex multi-objective control with high control accuracy and fast speed. For small MGs of small scale, centralized control has apparent advantages. However, with the development of the MG scale, distant DGs face the problem of interference and attenuation of signal transmission.

The increased number of DGs leads to a heavy communication burden and significant computation of the central controller. In addition, a single point of failure of the central controller can lead to global control paralysis and low reliability.

Decentralized control does not need communication and only relies on state information for regulation [27–30]. The decentralized control is a type of peer-to-peer control characterized by the fact that it does not require communication and relies only on local information for frequency and voltage restoration, as well as power equalization. Standard methods are to design state observers to estimate the global information of the system or to use virtual impedance to compensate for the effect of line impedance on power distribution. Decentralized control, being locally controlled and not dependent on communication, has a fast response time and high reliability. However, due to the lack of accurate global information, it usually has low control accuracy, poor anti-interference However, due to the lack of accurate global information, it usually has low control accuracy, poor interference immunity, quickly causes stability problems and is difficult to achieve complex global control objectives.

Distributed control uses communication between neighboring DGs to unite global information through mutual information exchange among DGs [31–42]. Distributed control also uses peer-to-peer control mode, using a distributed local controller instead of a central controller, avoiding the single point of failure of a centralized controller. The theoretical basis is a MAS consensus algorithm, where each DG communicates only with its neighboring nodes to "indirectly" share the global information of the system. It is ideal for large-scale and complex networks and meets the needs of a highly reliable MG. Distributed control has become the primary implementation method for secondary control of MGs.

1.2.3 Tertiary Control

The tertiary control is the energy management layer, which is responsible for developing the MG operation plan based on the distributed power output prediction, energy storage operation status, and load demand to maximize renewable energy utilization and the economical and safe operation of the MG. The third level of control is the optimization of MG operation. The specific energy management objectives are the minimum operating cost of DG, minimum system network loss, maximum utilization of renewable energy, minimum probability of power outage, minimum pollution emission, etc., using mixed integer linear programming method, particle swarm optimization algorithm, genetic algorithm and other solving methods. The third layer of control can develop generation plans for MG operation and instructions for lower layer control, such as reference values for interconnection power between MG operating in grid-connected process and external network, reference values for interconnection power within AC-DC hybrid MG, and generation plans for master control units within the independent MG. The tertiary layer control has a longer time scale than the first and secondary layer control and mainly adopts a centralized control method.

1.3 Multi-agent Control

With the rapid development of Internet technology and artificial intelligence, MASs, an essential branch of distributed artificial intelligence, play an important role in the control theory system and have been widely studied [43–52]. MAS is the consensus problem of a large-scale complex system composed of interconnected distributed autonomous or semi-autonomous subsystems. In other words, each agent in the system transmits information, interacts with and influences each other according to some connection, and thus the states of MASs converge. Due to the rapid development of computer technology, artificial intelligence theory, control theory, Internet of Things technology, etc., MASs have become a leading research focus in many disciplines [53, 54]. Many agents form a MAS, where individuals control their behaviour based on information from previous interactions with neighbouring individuals so that the system's overall behaviour moves toward global optimality. Thus the whole system accomplishes some control goal.

The general linear MAS dynamics containing N agents are represented as follows:

$$\dot{x}_i(t) = Ax_i(t) + Bu_i(t), \quad i = 1, ..., N. \tag{1.1}$$

where $x_i(t) \in \mathbb{R}^n$ and $u_i(t) \in \mathbb{R}^m$ denote the state and control input of the agent, respectively. $A \in \mathbb{R}^{n\times n}$ and $B \in \mathbb{R}^{m\times m}$ represents known matrix coefficients, and (A, B) is stabilizable.

The main control objective of the consensus problem is to make the following equation hold in any initial state $x_i(t_0) \in \mathbb{R}^n$ by designing suitable control inputs $u_i(t)$:

$$\lim_{t\to+\infty} \|x_i(t) - x_j(t)\| = 0, \quad i, j = 1, ..., N. \tag{1.2}$$

For linear MAS dynamics model (1.1), the following classical consensus control algorithm is usually considered [55]:

$$u_i(t) = \mu K \sum_{j=1}^{N} a_{ij}\left(x_i(t) - x_j(t)\right). \tag{1.3}$$

where μ is the custom coupling coefficient, K is the control gain matrix calculated from the algebraic Riccati equation, and a_{ij} represents the neighbor matrix coefficients, $a_{ij} = 1$ when agent i can receive information from intelligence j and vice versa $a_{ij} = 0$.

Bringing Eq. (1.3) into Eq. (1.1), the following compact form can be obtained:

$$\dot{x}(t) = (I_N \otimes A + \mu\mathcal{L} \otimes BK)x(t). \tag{1.4}$$

where $x(t) = [x_1^T, ..., x_N^T(t)]^T$. I_N are N-dimensional unit arrays, $\mathcal{L}$ represents the Laplace matrix formed by the multi-agent communication topology, and $\otimes$ represents the Kronecker product.

Note that in a previous study [56, 57], the design of the coupling weights μ relied on the eigenvalues of the Laplacian matrix $\mathcal{L}$. Thus the distributed control protocol (1.3) relies on global information. However, in large-scale MASs, obtaining global information takes much work. To address this problem, Li et al. [58] proposed a fully distributed consensus controller that does not require knowledge of the global information associated with the communication topology. The basic design idea is to decompose the controller parameters into local and neighboring variables, thus avoiding the dependence on global information. The advantage of this approach is that it is more general and can be applied to a broader range of situations but requires a more complex controller design. The following shows the fully distributed control protocol proposed in the literature [58]:

$$\begin{aligned} u_i(t) &= v_i(t) f_i\left(\xi_i^T(t) P^{-1} \xi_i(t) K \xi_i(t)\right) \\ v_i(t) &= \xi_i^T(t) \Phi \xi_i(t) \\ \xi_i(t) &= \sum_{j=1}^{N} a_{ij}(x_i(t) - x_j(t)). \end{aligned} \tag{1.5}$$

where $v_i(t)$ is the adaptive parameter of the ith agent to eliminate the effect of global information on consensus; Φ is a positive definite symmetric matrix calculated from the algebraic Riccati equation; and $\xi_i(t)$ represents the consensus error of the ith agent. This type of controller has the advantage of being independent of global information and can cope with general linear MASs. Currently, the use of adaptive techniques has become the primary means to eliminate the influence of global information on the consensus of MAS [59, 60].

1.3.1 Fault-Tolerant Control

As control systems grow in size and perform more complex tasks, various types of failures will inevitably occur in agents and communication networks. At the same time, faults can significantly affect a MAS's stability and cooperative control performance or even lead to the collapse of the whole system. Therefore, studying the cooperative fault-tolerant control of MASs when faults occur has become a key topic of interest for scholars in recent years. A well-tolerated fault control solution eliminates the adverse effects of failures on the entire system by triggering effective defence mechanisms. The problem of component fault brings complexity and difficulty to controlling system research and analysis design. Fault-tolerant control is a way to solve the control problem after a system fault [61, 62]. When a control system component fails, the designed controller can still make the closed-loop system operate stably and with the specified performance. They investigated the problem of fault-tolerant tracking control of linear and Lipschitz nonlinear MASs under actuator faults and leading bounded unknown inputs [63]. Chen et al. [64], Deng [65] investigates the fault-tolerant consensus problem of fully distributed MASs with constant multiplicative faults based on general undirected topology and directed topology, respectively. An adaptive fuzzy fault-tolerant control problem for a class of uncertain multiple-input multiple-output nonlinear systems with unknown control direction under time-varying asymmetric output constraints is investigated in [66]. Their communication network is considered a directed switching graph. Additionally, since one-way communication is more common in many communication networks, many studies are considering solving the problem of fully distributed fault-tolerant control based on directed graphs. In [67], the authors design a collaborative fault-tolerant control algorithm mainly for heterogeneous faults, such as failures and unknown biases that may exist in the unit actuators. Deng and Yang [68], Deng and Che [69] proposed solutions for the fault-tolerant control problem under MAS bias faults based on the directed fixed/switching network topology, respectively. However, the fault parameters of its PLOE faults are constant. In [70], a distributed adaptive event-triggered fault-tolerant controller is designed for time-varying PLOE fault parameters. However, the approach assumes knowledge of the upper and lower bounds of the fault parameters, which may only sometimes be practical. Considering the actuator fault problem, [71, 72] achieved global synchronization by constructing distributed reference observers based on neighborhood state information and decentralized adaptive fault-tolerant tracking controllers based on local estimation. In contrast, the distributed cooperative secondary control algorithm requires the structure of the communication graph and the number of total agents, which belong to the global information of the network communication topology [58, 73]. However, in practice, the minimum nonzero eigenvalue of the Laplace matrix for each agent is unknown. It should be noted that the fault-tolerant control results from [71] are not implemented in a fully distributed framework. However, for practical MG systems, it is impractical for each DG to compute the minimum eigenvalue of the communication network Laplace matrix or the scale of the DG. Designing a fully distributed control protocol with unknown fault parameters and realising complex decoupling between time-varying PLOE faults and adaptive controller gains in directed communication networks is more challenging, and it is still an open question.

In electrical and electronic systems, various faults in the control system are inevitable. Scholars have recently proposed some types of distributed secondary control strategies based on communication, actuator, and/or sensor faults in MG systems [74–76]. In [77], a stochastic consensus-based secondary restoration scheme for islanded MGs with communication delays and noise is proposed. Scholars have developed a resilient control framework that integrates an acknowledgment-based attack detection strategy and a communication restoration mechanism for the first time[78]. Actuators in DG perform control functions according to the nature of the deviation signal so that the controlled quantity is changed according to the desired value [79–81]. Practically, the actuator is the interface between the controller and the plant. In [82], fault-tolerant secondary voltage and frequency restoration methods for islanded MGs consider disturbances and actuator faults. In [83], the fault-tolerant control schemes are designed to handle island AC MGs under potential faults of sensors and actuators, modeled as additive faults. In [84], a consensus-based scheme is proposed to synchronizes the islanded MG's voltage and frequency to their nominal values for all DGs within finite time, irrespective of saturation and multiple faults, including partial loss of effectiveness and stuck faults simultaneously. The reliable functioning of the MG is maintained under parameter variations, noises and fault events (e.g., generator output outages). Especially for the latter case, [85] provides an adaptive control to handle fault events via subsequent control reconfiguration. The secondary consensus based controller for current sharing in DC MGs subjected to sensor fault is discussed in [86]. In [87], the impact from the multiplicative fault term and biased fault term of actuator faults are compensated by two adaptive dynamics of the distributed event-triggered fixed-time fault-tolerant secondary control, which does not require the information of the upper bound of the fault terms in advance. The work in [88] develops a novel fault-tolerant predictive controller for the class of DC MGs with nonlinearities, sensor and actuator faults, and noisy measurements. The main novelty of this work is presenting a novel dual-Extended Kalman filter based on which the system's noise-free states and sensor and actuator faults are estimated. The proposed control in [89] can accurately estimate the error (if there is any) in distributed energy resources voltage and current measurements in an islanded MG using sliding-mode observer-based fault diagnosis theory. This enables rectification of the erroneous sensor measurements and facilitates sliding fault-tolerant controller operation of the distributed energy resources units in an islanded MG unperturbed by the sensor faults. The adaptive fault-tolerant controller proposed in [90] is cleverly constructed and designed so that the closed-loop system is asymptotically stable regarding fault-tolerant output regulation. The distributed event-triggered fixed-time fault-tolerant secondary control framework in [91] can improve the power quality of the faulty AC MG by achieving the fixed-time convergence speed under the unknown actuator and sensor faults/attacks. The proposed method in [92] ensures accurate power sharing amongst inverters even in extreme operating conditions such as communication-link failure and sudden loss/turning OFF/ON some units. So, it supports plug-and-play features.

1.3.2 Event-Triggered Control

The control strategies described above are all implemented under time-triggered settings. Unlike time-triggered control, where data is transmitted on a fixed schedule, the event-triggered approach samples information when a pre-specified event is triggered [93–102]. As a result, the communication cost and the amount of transmitted data will be significantly reduced. Using event-triggered techniques, a class of nonlinear MASs is studied for the distributed control problem in [103]. A new dynamic event triggering mechanism is proposed, guaranteeing a larger event interval than the static event triggering mechanism [104]. Because of the reliance on combinatorial measurement errors in the triggering mechanism, agents still require continuous access to neighboring states [103, 104]. Dynamic event-triggered consistent control of general linear masses is investigated while avoiding the problem of continuous communication agents in [105]. However, [103–105] generally requires the determination of some parametric control laws or trigger functions dependent on the nonzero eigenvalues of the Laplace operator matrix of the global information network with the communication correlation graph. Therefore, the event-based consensus protocols in the above works must still be fully distributed. Further, [59, 60, 106–109] relies only on local information and implements a distributed event-triggering algorithm that avoids continuous communication. The work in [109] solved the problem of unknown upper and lower bounds of fault factors caused by time-varying multiplicative and additive faults. Distributed adaptive event-triggered protocols based on locally sampled state information are designed in [59, 106]. Distributed event-triggered protocols that avoid continuous communication were implemented based on the symmetric topology of undirected graphs in [107]. Fully distributed event-triggered control algorithms based on directed graphs were proposed in [60, 108], and the algorithms rely on multiple and improved individual event triggering mechanisms, respectively. More practical fault-tolerant strategies for overcoming this problem of actuator fault in an event-triggered framework have recently emerged. Furthermore, a two-channel event-triggered adaptive fault-tolerant control approach was designed to handle nonlinear systems with disturbances [110], and a fully distributed dynamic event-triggered fault-tolerant control problem based on heterogeneous linear MAS was discussed in [111]. Although continuous update is not required in the controller designs of the above methods [110, 111], continued access to neighbor states is still required in trigger-error or trigger-threshold designs. A consensus error prediction model was proposed in [70, 112] to avoid continuous communication between DGs. However, it may increase the computational burden, and the method will only be applicable in cases where full-state measurements are available. Event-triggered fault-tolerant protocols that do not require continuous communication to implement controller updates and trigger state detection were proposed in [113]. However, the protocol in [113] can only be implemented when the multiplicative faults of each agent are the same and constant. Unknown time-varying heterogeneous actuator multiplicative faults and nonlinear dynamics make designing event-triggered fault-tolerant control protocols based on discrete communication challenging. A fully distributed directed graph control method [114] with multiple event-triggered mechanisms was proposed in the ideal state of the actuator. Moreover, the trigger mechanism with a strictly positive minimum inter-event times (MIET) is a more critical requirement than excluding Zeno phenomena [115]. A uniformly strictly positive minimal event interval time is sufficient to exclude Zeno behavior, the phenomenon of infinitely many events occurring in a finite time interval. It is also more practical than traditional event triggering because the hardware constraint always imposes a positive minimum time constraint between two consecutive event times. To avoid Zeno behavior while ensuring the presence of positive MIET.

Event-based control methods are more suitable for electric power systems due to the low number of updates, communication, and computation [28, 116–121]. With this context in mind, event-triggered strategies have emerged in secondary voltage control. A distributed periodic event-triggered optimal control scheme is proposed in [122] for minimizing the cost of DC MG generation and bus average voltage regulation. One of the discrete-time periodic event triggering mechanisms leads to a significant reduction in communication requirements. In order to achieve proportional load current sharing and bus average voltage regulation in DC MGs, a discrete-time algorithm for periodic event triggering that requires only local and neighboring trigger information is proposed in [123]. In [124], using the aperiodicity and intermittent properties of the communication, a simple detection principle is proposed to detect and isolate the compromised communication links in a timely and economical fashion, which improves the resilience of the a system against attacks. In [125], a hybrid ETM is proposed to reduce the utilization of communication resources, which imposes a lower bound on the event-triggering intervals such that Zeno behavior is excluded. In [126], the proposed dynamic mechanism guarantees the existence of a designable positive MIET that is independent of the state trajectories and adjustable by the control parameters. Authors in [127] propose an event-based secondary voltage restoration scheme for delay-affected MGs in communication topologies. In [128], a distributed secondary control scheme is proposed with an event-triggered communication mechanism based on sampled data. The event-triggered hybrid control approach proposed in [129] to ensure high security in hybrid energy generation systems exhibits a remarkable superiority in addressing event-triggered issues about power systems. Moreover, a distributed event-triggered control scheme was proposed for accurate voltage restoration and precise current sharing in islanded DC MGs [130], while distributed event-triggered laws were introduced for secondary frequency and voltage restoration in delay-affected AC MGs [127, 131]. Fully distributed event-triggered fault-tolerant voltage regulation strategies were proposed using the combined sample measurement error method [132, 133].

1.4 Publication Outline

The general layout of this monograph is as follows. A class of cooperative control strategies based on undirected graphs is given under the output voltage regulation problem. It is further extended to directed communication networks. Then the load voltage consistency regulation problem is based on asymmetric communication networks. Then PCC voltage regulation based on containment control is investigated. Finally, the results are extended to the voltage regulation at the PCC end of multi-MG. The main contents of this monograph are as follows:

This chapter presents the research background, motivations and research problems, which involve basic structure of MG, hierarchical control structures, multi-agent control and then the outline of the monograph is listed.

Chapter 2 investigates the problem of secondary voltage restoration in an islanded MG, in which the actuator of the DGs may coexist with partial loss of effectiveness (PLOE) fault and bias fault. For each DG, an adaptive event-triggered fault-tolerant

(ETFT) control protocol is designed to compensate for the effects of actuator faults in the DGs, thereby restoring the voltage to the reference value. The dependent event triggering mechanism saves the processor's computational resources. A desirable feature of the protocol proposed in this chapter is that the control protocol relies only on relative information between neighboring DG's, independent of global information about the network graph, fault boundaries, and network scale. It means that the protocol is implemented in a fully distributed framework. The protocol fully applies to the common ETFT consensus control problem of linear MASs with actuator faults. Furthermore, comprehensive theoretical arguments for consensus stability and analysis of Zeno behavior ensure the approach's feasibility. The simulation results verify the effectiveness of the algorithm.

Chapter 3 investigates an ETFT control strategy for voltage restoration in MGs. The presence of PLOE and bias faults in the actuators are considered. The proposed algorithm is independent of the global information related to the directed communication network and does not involve the boundaries of the fault factors. Each DG only needs to know the relative information between its neighbors, which means that the algorithm is fully distributed. The boundary layer technique is used to construct smooth ETFT controllers, avoiding the undesirable chattering effect of discontinuous controllers. Furthermore, a novel consensus error prediction model is developed to eliminate the need for continuous communication between DGs. The proposed algorithm is not only applicable to the secondary voltage restoration of MGs but can also be extended to general linear MASs. Finally, the simulation results verify the effectiveness of the algorithm.

Chapter 4 proposes an observer-based event-driven fault-tolerant (OBED FT) secondary control strategy for AC MGs to achieve load voltage regulation. First, the input-output feedback linearization method transforms the voltage regulation issue into an output feedback tracking problem for linear MASs with nonlinear dynamics. This transformation provides the necessary pre-processing for load voltage regulation. Then, an OBEDFT secondary control protocol that considers full-state immeasurability is proposed. The actuators of DGs may experience PLOE and bias faults, and these fault parameters may be heterogeneous and time-varying. The protocol introduces adaptive techniques to avoid information related to fault parameters while using event-driven mechanisms to achieve discrete measurements of neighboring DG. Additionally, the protocol uses boundary layers to construct smooth controllers that prevent the chattering effect caused by non-smooth controllers. Finally, simulation results confirm the effectiveness of this load voltage regulation strategy.

Chapter 5 presents a distributed containment-based load voltage and consensus-based frequency fault-tolerant secondary control strategy. The strategy considers distributed generators (DGs) that communicate through a directed network whose fault models include both actuators- partial loss of effectiveness (PLOE) and deviation faults. First, we introduce the idea of containment control in the load voltage secondary control strategy. Unlike the traditional consensus-based output voltage secondary control strategy, this strategy limits the load voltage on each bus to a reasonable range while having a voltage difference between the loads to allow power flow. Next, we use a hierarchical mechanism to decouple the distributed secondary control problem into a cooperative control problem for virtual systems and a decentralized control problem for the existing system. Note that the strategy avoids using global information about the directed communication network and quantitative information about the size, frequency, and type of actuator faults. In addition, we design a consensus-based frequency regulation and power-sharing control strategy. Finally, simulation results verify the effectiveness of the process.

Chapter 6 proposes a distributed containment-based secondary voltage control strategy with positive MIET. Firstly, the feedback linearization technique allows us to deal with nonlinear DG dynamics using a reduced-order approach. The critical bus voltage regulation problem of an AC MG is transformed into an output feedback tracking problem of a linear MAS containing nonlinear dynamics. This process is a necessary pre-processing process in the design of control algorithms. In addition, an event-triggered strategy based on containment control is proposed. The strategy limits each critical bus voltage within a reasonable range while having a voltage difference between sub-MGs to allow power flow. Furthermore, an innovative trigger mechanism has been designed based on multiple measurement errors. Based on this mechanism, the controllers are updated in an acyclic manner at the moment of event sampling, saving computational resources and transmission load. In addition, the proposed event-triggering mechanism guarantees the absence of Zeno behavior and a strictly positive MIET between each DG event. It is also more practical than the traditional event triggering because the hardware constraint always imposes a positive minimum time constraint between two consecutive event times. The protocol proposed in this chapter is fully distributed and scalable and does not depend on global information about the network graph. Finally, simulation results verify the effectiveness of the strategy.

References

1. Hosseini, S., Abyaneh, H., Sadeghi, S., et al.: An overview of microgrid protection methods and the factors involved. Renew. Sustain. Energy Rev. **64**, 174–186 (2016)
2. Herrera, L., Zhang, W., Wang, J.: Stability analysis and controller design of DC microgrids with constant power loads. IEEE Trans. Smart Grid **5**(5), 2543–2552 (2014)
3. Mahmood, H., Michaelson, D., Jiang, J.: Accurate reactive power sharing in an islanded microgrid using adaptive virtual impedances. IEEE Tans. Power Electron. **30**(3), 1605–1617 (2015)
4. Adeyanju, O.M. ,Siano, P., Canha, L.N.: Dedicated microgrid planning and operation approach for distribution network support with pumped-hydro storage. IEEE Trans. Ind. Inf. https://doi.org/10.1109/TII.2022.3216299
5. Manojkumar, R., Kumar, C., Ganguly, S., Gooi, H.B., Mekhilef, S., Catalo, J.P.S.: Rule-based peak shaving using master-slave level optimization in a diesel generator supplied microgrid. IEEE Trans. Power Syst. https://doi.org/10.1109/TPWRS.2022.3187069
6. Yaqoob, M., Lashab, A., Vasquez, J.C., Guerrero, J.M., Orchard, M.E., Bintoudi, A.D.: A comprehensive review on small satellite microgrids. IEEE Trans. Power Electron. **37**(10), 12741–12762 (2022)

7. Neto, P.J.D.S. , Barros, T.A.D.S., Silveira, J.P.C., Filho, E.R., Vasquez, J.C., Guerrero, J.M.: Power management strategy based on virtual inertia for DC microgrids. IEEE Trans. Power Electron. **35**(11), 12472–12485 (2020)
8. Harmon, E., Ozgur, U., Cintuglu, M.H., de Azevedo, R., Akkaya, K., Mohammed, O.A.: The internet of microgrids: a cloud-based framework for wide area networked microgrids. IEEE Trans. Ind. Inf. **14**(3), 1262–1274 (2018)
9. Reddy, O.Y.,Jithendranath J., Chakraborty, A.K., Guerrero, J.M.: Stability constrained optimal operation of standalone DC microgrids considering load and solar PV uncertainties. IEEE Trans. Power Deliv. https://doi.org/10.1109/TPWRD.2023.3253623
10. Mishra, A., Jena, P.: Application of blockchain technology for microgrid restoration. IEEE Trans. Power Deliv. https://doi.org/10.1109/TPWRD.2022.3226659
11. Liu, L., Yang, G.H.: Distributed optimal economic environmental dispatch for microgrids over time-varying directed communication graph. IEEE Trans. Netw. Sci. Eng. **8**(2), 1913–1924 (2021)
12. Huang, B., Liu, L., Zhang, H., et al.: Distributed optimal economic dispatch for microgrids considering communication delays. IEEE Trans. Syst. Man Cybern. 1634–1642 (2019)
13. Wu, X., Xu, Y., He, J., Wang, X., Vasquez, J., Guerrero, J.: Pinning-based hierarchical and distributed cooperative control for AC microgrid clusters. IEEE Trans. Power Electron. **35**(9), 9865–9885 (2020)
14. Li, Z., Cheng, Z., Liang, J., et al.: Distributed event-triggered secondary control for economic dispatch and frequency restoration control of droop-controlled AC microgrids. IEEE Trans. Sustain. Energy **11**(3), 1938–1950 (2020)
15. Ma, W., Wang, J., Gupta, V., et al.: Distributed energy management for networked microgrids using online ADMM with regret. IEEE Trans. Smart Grid **9**(2), 847–856 (2018)
16. Azevedo, R.D., Cintuglu, M.H., Ma, T., Mohammed, O.: Multiagent-based optimal microgrid control using fully distributed diffusion strategy. IEEE Trans. Smart Grid **8**(4), 1997–2008 (2017)
17. Che, L., Shahidehpour, M., Alabdulwahab, A., Al-Turki, Y.: Hierarchical coordination of a community microgrid with AC and DC microgrids. IEEE Trans. Smart Grid **6**(6), 3042–3051 (2015)
18. Fan, B., Guo, S., Peng, J., Yang, Q., Liu, W., Liu, L.: A consensus-based algorithm for power sharing and voltage regulation in DC microgrids. IEEE Trans. Ind. Inf. **16**(6), 3987–3996 (2020)
19. Chen, Z., Yu, X., Xu, W., Wen, G.: Modeling and control of islanded DC microgrid clusters with hierarchical event-triggered consensus algorithm. IEEE Trans. Circ. Syst I: Regular Papers **68**(1), 376–386 (2021)
20. Guerrero, J.M., Vasquez, J.C., Matas, J., de Vicuna, L.G., Castilla, M.: Hierarchical control of droop-controlled AC and DC microgrids-a general approach toward standardization. IEEE Trans. Ind. Electron. **58**(1), 158–172 (2011)
21. Morstyn, T., Hredzak, B., Agelidis, V.: Control strategies for microgrids with distributed energy storage systems: an overview. IEEE Trans. Smart Grid **9**(4), 3652–3666 (2018)
22. Chen, X., Shi, M., Sun, H., et al.: Distributed cooperative control and stability analysis of multiple DC electric springs in a DC microgrid. IEEE Trans. Ind. Electron. **65**(7), 5611–5622 (2018)
23. Shafiee, Q., Nasirian, V., Vasquez, L., et al.: A multi-functional fully distributed control framework for AC microgrids. IEEE Trans. Smart Grid **9**(4), 3247–3258 (2018)
24. Chen, X., Shi, M., Zhou, J., Chen, Y., Zuo, W., Wen, J., He, H.: Distributed cooperative control of multiple hybrid energy storage systems in a DC microgrid using consensus protocol. IEEE Trans. Ind. Electron. **67**(3), 1968–1979 (2020)
25. Chen, G., Zhao, Z.: Delay effects on consensus-based distributed economic dispatch algorithm in microgrid. IEEE. Trans. Power Syst. **33**(1), 602–612 (2018)
26. Qian, T., Liu, Y., Zhang, W., Tang, W., Shahidehpour, M.: Event-triggered updating method in centralized and distributed secondary controls for islanded microgrid restoration. IEEE Trans. Smart Grid **11**(2), 1387–1395 (2020)

27. Gu, W., Lou, G., Tan, W., et al.: A nonlinear state estimator-based decentralized secondary voltage control scheme for autonomous microgrids. IEEE Trans. Power Syst. **32**(6), 4794–4804 (2017)
28. Chen, M., Xiao, X., Guerrero, J.M.: Secondary restoration control of islanded microgrids with a decentralized event-triggered strategy. IEEE Trans. Ind. Inf. **14**(9), 3870–3880 (2018)
29. Wan, K., Chen, Y., Zhao J. , Yu, M.: A decentralized fault-tolerant control for DC microgrids against disturbances and actuator faults. IEEE Trans Smart Grid
30. Wang, H. , Yu, Y., Jiang, Y.: Fully decentralized multiagent communication via causal inference. IEEE Trans. Neural Netw. Learn. Syst. https://doi.org/10.1109/TNNLS.2022.3165114
31. Lu, X., Yu, X., Lai, L., Guerrero, J., Zhou, H.: Distributed secondary voltage and frequency control for islanded microgrids with uncertain communication links. IEEE Trans. Ind. Inf. **13**(2), 448–460 (2017)
32. Dehkordi, N.M., Sadati, N., Hamzeh, M.: Distributed robust finite-time secondary voltage and frequency control of islanded microgrids. IEEE Trans. Power Syst. **32**(5), 3648–3659 (2017)
33. Bidram, A., Davoudi, A., Lewis, F.L., Guerrero, J.M.: Distributed cooperative secondary control of microgrids using feedback linearization. IEEE Trans. Power Syst. **28**(3), 3462–3470 (2013)
34. Bidram, A., Davoudi, A., Lewis, F.L., Ge, S.S.: Distributed adaptive voltage control of inverter-based microgrids. IEEE Trans. Energy Convers. **29**(4), 862–872 (2014)
35. Ning, B., Han, Q.L., Ding, L.: Distributed finite-time secondary frequency and voltage control for islanded microgrids with communication delays and switching topologies. IEEE Trans. Cybern. **51**(8), 3988–3999 (2021)
36. Yang, H. , Li, T. ,Long, Y. , Xiao, Y.: Event-triggered distributed secondary control with model-free predictive compensation in AC/DC networked microgrids under DoS attacks. IEEE Trans. Cybern. (2022)
37. Guo, F., Wen, C., Mao, J., Song, Y.D.: Distributed secondary voltage and frequency restoration control of droop-controlled inverter-based microgrids. IEEE Trans. Ind. Electron. **62**(7), 4355–4364 (2015)
38. Dehkordi, N.M., Sadati, N., Hamzeh, M.: Fully distributed cooperative secondary frequency and voltage control of islanded microgrids. IEEE Trans. Energy Convers. **32**(2), 675–685 (2017)
39. Ge, P., Dou, X., Quan, X., Hu, Q., Sheng, W., Wu, Z., et al.: Extended-state-observer-based distributed robust secondary voltage and frequency control for an autonomous microgrid. IEEE Trans. Sustain. Energy **11**(1), 195–205 (2020)
40. Li, X., Wen, C., Chen, C., Xu, Q.: Adaptive resilient secondary control for microgrids with communication faults. IEEE Trans. Cybern. **52**(8), 8493–8503 (2022)
41. Dehkordi, N.M., Baghaee, H.R., Sadati, N., Guerrero, J.M.: Distributed noise-resilient secondary voltage and frequency control for islanded microgrids. IEEE Trans. Smart Grid **10**(4), 3780–3790 (2019)
42. Guo, X.G., Liu, P.M., Wang, J.L., Ahn, C.K.: Event-triggered adaptive fault-tolerant pinning control for cluster consensus of heterogeneous nonlinear multi-agent systems under aperiodic dos attacks. IEEE Trans. Netw. Sci. Eng. **8**(2), 1941–1956 (2021)
43. Dou, C., Yue, D., Li, X., Xue, Y.: Mas-based management and control strategies for integrated hybrid energy system. IEEE Trans. Ind. Inf. **12**(4), 1332–1349 (2016)
44. Bui, V., Hussain, A., Kim, H.: A multiagent-based hierarchical energy management strategy for multi-microgrids considering adjustable power and demand response. IEEE Trans. Smart Grid **9**(2), 1323–1333 (2018)
45. Yue, D. , He, Z., Dou, C.: Cloud-edge collaboration based distribution network reconfiguration for voltage preventive control. IEEE Trans. Ind. Inf. https://doi.org/10.1109/TII.2023.3247028
46. Yue, D., Zhang, H.F., Dou, C.X.: Cooperative optimal control of hybrid energy systems. Springer Nature Singapore Pte Ltd (2021)

47. Zhao, Y., Xian, C., Wen, G., Huang, P., Ren, W.: Design of distributed event-triggered average tracking algorithms for homogeneous and heterogeneous multiagent systems. IEEE Trans. Autom. Control **67**(3), 1269–1284 (2022)
48. Ren, W., Atkins, E.: Distributed multi-vehicle coordinated control via local information exchange. Int. J. Robust Nonlinear Control: IFAC-Affiliated Journal **17**(10–11), 1002–1033 (2007)
49. Xiao, F., Wang, L., Gao, J.: Finite-time formation control for multi-agent systems. Automatica **45**(11), 2605–2611 (2009)
50. Hu, G.: Robust consensus tracking of a class of second-order multi-agent dynamic systems. Syst. Control Lett. **61**(1), 134–142 (2012)
51. Meng, X., Chen, T.: Event based agreement protocols for multi-agent networks. Automatica **49**(7), 2125–2132 (2013)
52. Du, H., He, Y., Cheng, Y.: Finite-time synchronization of a class of second-order nonlinear multi-agent systems using output feedback control. IEEE Trans. Circ. Syst. I: Regular Papers **61**(6), 1778–1788 (2014)
53. Jie, M., Wei, R., Jie, C.: Distributed consensus of second-order multi-agent systems with heterogeneous unknown inertias and control gains under a directed graph. IEEE Trans. Autom. Control **61**(8), 2019–2034 (2016)
54. Liu, D., Shen, M., Jing, Y., Wang, Q.-G.: Distributed optimization of nonlinear multiagent systems via event-triggered communication. IEEE Trans. Circ. Syst. II: Express Briefs https://doi.org/10.1109/TCSII.2022.3225800
55. Li, Z., Duan, Z.: Cooperative control of multi-agent systems: a consensus region approach. CRC Press, Boca Raton, FL, USA (2014)
56. Li, Z., Duan, Z., Chen, G., et al.: Consensus of multiagent systems and synchronization of complex networks: Aunified viewpoint. IEEE Trans. Circ. Syst. I: Regular Papers **57**(1), 213–224 (2010)
57. Lewis, F., Zhang, H., Hengster-Movric, K., et al.: Cooperative control of multi-agent systems: optimal and adaptive design approaches. Springer Publishing Company, Incorporated (2014)
58. Li, Z., Wen, G., Duan, Z., Ren, W.: Designing fully distributed consensus protocols for linear multi-agent systems with directed graphs. IEEE Trans. Autom. Control. **60**(4), 1152–1157 (2015)
59. Cheng, B., Li, Z.: Fully distributed event-triggered protocols for linear multi-agent networks. IEEE Trans. Autom. Control **64**(4), 1655–1662 (2019)
60. Li, X., Sun, Z., Tang, Y., Karimi, H.R.: Adaptive event-triggered consensus of multiagent systems on directed graphs. IEEE Trans. Autom. Control **66**(4), 1670–1685 (2021)
61. Sun, J., Zhang, H., Wang, Y., Sun, S.: Fault-tolerant control for stochastic switched IT2 fuzzy uncertain time-delayed nonlinear systems. IEEE Trans. Cybern. **52**(2), 1335–1346 (2022)
62. Allerhand, L.I., Shaked, U.: Robust switching-based fault tolerant control. IEEE Trans. Autom. Control **60**(8), 2272–2276 (2015)
63. Zuo, Z., Zhang, J., Wang, Y.: Adaptive fault-tolerant tracking control for linear and lipschitz nonlinear multi-agent systems. IEEE Trans. Ind. Electron. **62**(6), 3923–3931 (2015)
64. Chen, S., Ho, D., Li, L., Liu, M.: Fault-tolerant consensus of multi-agent system with distributed adaptive protocol. IEEE Trans. Cybern. **45**(10), 2142–2155 (2017)
65. Deng, C.: Cooperative fault-tolerant output regulation of linear heterogeneous multiagent systems under directed network topology. IEEE Trans. Syst. Man Cybern. Syst. **51**(8), 4773–4781 (2021)
66. Ruan, Z., Yang, Q., Ge, S.S., Sun, Y.: Adaptive fuzzy fault tolerant control of uncertain MIMO nonlinear systems with output constraints and unknown control directions. IEEE Trans. Fuzzy Syst. **30**(5), 1224–1238 (2022)
67. Afshari, A., Karrari, M., Baghaee, H.R., Gharehpetian, G.B., Karrari, S.: Cooperative fault-tolerant control of microgrids under switching communication topology. IEEE Trans. Smart Grid **11**(3), 1866–1879 (2020)
68. Deng, C., Yang, G.H.: Distributed adaptive fuzzy control for nonlinear multi-agent systems under directed graphs. IEEE Trans. Fuzzy Syst. **26**(99), 1356–1366 (2018)

69. Deng, C., Che, W.W.: Fault-tolerant fuzzy formation control for a class of nonlinear multiagent systems under directed and switching topology. IEEE Trans. Syst Man Cybern. Syst. **51**(9), 5456–5465 (2021)
70. Chen, J., Chen, B., Zeng, Z.: Adaptive dynamic event-triggered fault-tolerant consensus for nonlinear multiagent systems with directed/undirected networks. IEEE Trans. Cybern. (2022)
71. Wang, X., Yang, G.: Fault-tolerant consensus tracking control for linear multiagent systems under switching directed network. IEEE Trans. Cybern. **50**(5), 1921–1930 (2020)
72. Xiao, S., Dong, J.: Cooperative fault-tolerant containment control for nonlinear multiagent systems with switching directed topologies based on hierarchical mechanism. IEEE Trans. Syst. Man Cybern. Syst. **52**(9), 5424–5433 (2022)
73. Song, Q., Wen, G., Yu, W., Meng, D., Lu, W.: Fully distributed synchronization of complex networks with adaptive coupling strengths. IEEE Trans. Cybern. **52**(11), 11581–11593 (2022)
74. Chen, C., Lewis, F.L., Xie, S., Modares, H., Liu, Z., Zuo, S., et al.: Resilient adaptive and $h\infty$ controls of multi-agent systems under sensor and actuator faults. Automatica **102**, 19–26 (2019)
75. Lai, J., Lu, X., Dong, Z., Cheng, S.: Resilient distributed multiagent control for AC microgrid networks subject to disturbances. IEEE Trans. Syst. Man Cybern. Syst. **52**(1), 43–53 (2022)
76. Ning, B., Han, Q.-L., Ding, L.: Distributed secondary control of AC microgrids with external disturbances and directed communication topologies: A full-order sliding-mode approach. IEEE/CAA J. Automatica Sinica **8**(3), 554-C564 (2021)
77. Shahab, M.A., Mozafari, S.B., Soleymani, S., Mahdian, N., Mohammadnezhad, H., Guerrero, J.M.: Stochastic consensus-based control of μgs with communication delays and noises. IEEE Trans. Power Syst. **34**(5), 3573–3581 (2019)
78. Ding, L., Han, Q.L., Ning, B., Yue, D.: Distributed resilient finite-time secondary control for heterogeneous battery energy storage systems under denial-of-service attacks. IEEE Trans. Ind. Inf. **16**(7), 4909–4919 (2020)
79. Hua, Y., Dong, X., Li, Q., Ren, Z.: Distributed fault-tolerant time-varying formation control for second-order multi-agent systems with actuator failures and directed topologies. IEEE Trans. Circ. Syst. II Express Briefs **65**(6), 774–778 (2018)
80. Khalili, M., Zhang, X., Polycarpou, M.M., Parisini, T., Cao, Y.: Distributed adaptive fault-tolerant control of uncertain multi-agent systems. Automatica **87**, 142–151 (2018)
81. Deng, C., Yang, G.H.: Distributed adaptive fault-tolerant containment control for a class of multi-agent systems with non-identical matching non-linear functions. IET Control Theory Appl. **10**(3), 273–281 (2016)
82. Shahab, M.A., Mozafari, B., Soleymani, S., Dehkordi, N.M., Shourkaei, H.M., Guerrero, J.M.: Distributed consensus-based fault tolerant control of islanded microgrids. IEEE Trans. Smart Grid **11**(1), 37–47 (2020)
83. Afshari, A., Karrari, M., Baghaee, H.R., Gharehpetian, G.B.: Distributed fault-tolerant voltage/frequency synchronization in autonomous AC microgrids. IEEE Trans. Power Syst. **35**(5), 3774–3789 (2020)
84. Jamali, M., Baghaee, H.R., Sadabadi, M.S., Gharehpetian, G.B., Anvari-Moghaddam, A., Blaabjerg, F.: Distributed finite-time fault-tolerant control of isolated AC microgrids considering input constraints. IEEE Trans. Smart Grid **13**(6), 4525–4537 (2022)
85. Prodan, I., Zio, E.: Stoican, F: Fault tolerant predictive control design for reliable microgrid energy management under uncertainties. Energy **91**, 20–34 (2015)
86. Huang, M., Ding, L., Li, W., Chen, C.-Y.: Liu, Z: Distributed observer-Based H_∞ fault-tolerant control for DC microgrids with sensor fault. IEEE Trans. Circ. Syst. I: Regular Papers **68**(4), 1659–1670 (2021)
87. Wang, Z., et al.: Distributed event-triggered fixed-time fault-tolerant secondary control of islanded AC microgrid. IEEE Trans. Power Syst. **37**(5), 4078–4093 (2022)
88. Vafamand, N., Arefi, M.M., Asemani, M.H., Javadi, M.S., Wang, F., Catalo, J.P.S.: Dual-EKF-based fault-tolerant predictive control of nonlinear DC microgrids with actuator and sensor faults. IEEE Trans. Ind. Appl. **58**(4), 5438–5446 (2022)

89. Saha, S., Gholami, S., Khan Prince, M.K.: Sensor fault-resilient control of electronically coupled distributed energy resources in islanded microgrids. IEEE Trans. Ind. Appl. **58**(1), 914–929 (2022)
90. Fan, Q.-Y., Deng, C., Ge, X., Wang, C.-C.: Distributed adaptive fault-tolerant control for heterogeneous multiagent systems with time-varying communication delays. IEEE Trans. Syst. Man Cybern. Syst. **52**(7), 4362–4372 (2022)
91. Wang, Z., et al.: A distributed event-triggered fixed-time fault-tolerant secondary control framework of islanded AC microgrid against faults and communication constraints. IEEE Trans. Power Syst. **37**(5), 3817–3833 (2022)
92. Mohammed, N., Lashab, A., Ciobotaru, M., Guerrero, J.M.: Accurate reactive power sharing strategy for droop-based islanded AC microgrids. IEEE Trans. Ind. Electron. **70**(3), 2696–2707 (2023)
93. Guo, X., Zhang, D., Wang, J., Ahn, C.K.: Adaptive memory event-triggered observer-based control for nonlinear multi-agent systems under dos attacks. IEEE/CAA J. Automatica Sinica **8**(10), 1644–1656 (2021)
94. Wang, F., Long, L.: Switched-observer-based event-triggered adaptive fuzzy funnel control for switched nonlinear systems. IEEE Trans. Fuzzy Syst. **30**(6), 1773–1787 (2022)
95. Li, Z., Ren, W., Liu, X., Fu, M.: Consensus of multi-agent systems with general linear and Lipschitz nonlinear dynamics using distributed adaptive protocols. IEEE Trans. Autom. Control **58**(7), 17861791 (2013)
96. Ge, X., Han, Q.L.: Distributed formation control of networked multi-agent systems using a dynamic event-triggered communication mechanism. IEEE Trans. Ind. Electron. **64**(10), 8118–8127 (2017)
97. Du, S., Liu, T., Ho, D.: Dynamic event-triggered control for leader-following consensus of multiagent systems. IEEE Trans. Syst. Man Cybern. Syst. **50**(9), 3243–3251 (2020)
98. Li, X., Tang, Y., Karimi, H.R.: Consensus of multi-agent systems via fully distributed event-triggered control. Automatica **116**, 108898 (2020)
99. Cao, L., Li, H., Dong, G., Lu, R.: Event-triggered control for multiagent systems with sensor faults and input saturation. IEEE Trans. Syst. Man Cybern. Syst. **51**(6), 3855–3866 (2021)
100. Wang, L., Dong, J.: Reset event-triggered adaptive fuzzy consensus for nonlinear fractional-order multiagent systems with actuator faults. IEEE Trans. Cybern. **53**(3), 1868–1879 (2023)
101. Du, C., Liu, X., Ren, W., Lu, P., Liu, H.: Finite-time consensus for linear multiagent systems via event-triggered strategy without continuous communication. IEEE Trans. Control Netw. Syst. **7**(1), 19–29 (2020)
102. Ruan, X., Feng, J., Xu, C., Wang, J.: Observer-based dynamic event-triggered strategies for leader-following consensus of multi-agent systems with disturbances. IEEE Trans. Netw. Sci. Eng. **7**(4), 3148–3158 (2020)
103. Li, Z., Yan, J., Yu, W., Qiu, J.: Event-triggered control for a class of nonlinear multiagent systems with directed graph. IEEE Trans. Syst. Man Cybern. Syst. **51**(11), 6986–6993 (2021)
104. Hu, W., Yang, C., Huang, T., Gui, W.: A distributed dynamic event-triggered control approach to consensus of linear multiagent systems with directed networks. IEEE Trans. Cybern. **50**(2), 869–874 (2020)
105. He, W., Xu, B., Han, Q., Qian, F.: Adaptive consensus control of linear multiagent systems with dynamic event-triggered strategies. IEEE Trans. Cybern. **50**(7), 2996–3008 (2020)
106. Cheng, B., Li, Z.: Designing fully distributed adaptive event-triggered controllers for networked linear systems with matched uncertainties. IEEE Trans. Neural Netw. Learn. Syst. **30**(12), 3645–3655 (2019)
107. Liu, S., Sun, J., Zhang, H., Zhai, M.: Fully distributed event-driven adaptive consensus of unknown linear systems. IEEE Trans. Neural Netw. Learn. Syst. (2022)
108. Zhai, M., Sun, Q., Wang, B., Wang, R., Liu, Z., Zhang, H.: Distributed secondary voltage control of microgrids with actuators bias faults and directed communication topologies: Event-triggered approaches. Int. J. Robust Nonlinear Control **32**(8), 4422–4437 (2022)
109. Luo, S., Ye, D.: Adaptive double event-triggered control for linear multi-agent systems with actuator faults. IEEE Trans. Circ. Syst. I: Regular Papers **66**(12), 4829–4839 (2019)

110. Guo, B., Chen, Y.: Event-triggered robust adaptive sliding mode fault-tolerant control for nonlinear systems. IEEE Trans. Ind. Inf. **16**(11), 6982–6992 (2020)
111. Guo, X.G., Liu, P.M., Wang, J.L., Ahn, C.K.: Event-triggered adaptive fault-tolerant pinning control for cluster consensus of heterogeneous nonlinear multi-agent systems under aperiodic dos attacks. IEEE Trans. Netw. Sci. Eng. **8**(2), 1941–1956 (2021)
112. Sun, J., Tan, Z., Liu, S. , Zhang, H., Chuo, W.: Fully distributed event-driven coordination with actuator faults. IEEE Trans Cybern (2022)
113. Xu, Y., Wu, Z.G.: Distributed adaptive event-triggered fault-tolerant synchronization for multiagent systems. IEEE Trans. Ind. Electron. **68**(2), 1537–1547 (2021)
114. Su, H., Cheng, B., Li, Z.: Fully distributed event-based protocols for lur'e systems over directed graphs. IEEE Trans. Circ. Syst. II Express Briefs **69**(3), 1812–1816 (2022)
115. Qian, Y.Y. , Wan, Y.: Design of distributed adaptive event-triggered consensus control strategies with positive minimum inter-event times. Automatica (2021)
116. Wang, Y., Deng, C., Liu, D., Xu, Y., Dai, J.: Unified real power sharing of generator and storage in islanded microgrid via distributed dynamic event-triggered control. IEEE Trans. Power Syst. **36**(3), 1713–1724 (2021)
117. Mohammadi, K., Azizi, E., Choi, J., Hamidi-Beheshti, M.-T., Bidram, A., Bolouki, S.: Asynchronous periodic distributed event-triggered voltage and frequency control of microgrids. IEEE Trans. Power Syst. **36**(5), 4524–4538 (2021)
118. Yu, C., Zhou, H., Lu, X., Lai, J., Liu, G.-P.: Distributed optimal synchronization rate control for AC microgrids under event-triggered mechanism. IEEE Trans. Power Syst. **36**(3), 1780–1793 (2021)
119. Wang, Y., Nguyen, T.L., Xu, Y., Li, Z., Tran, Q.-T., Caire, R.: Cyber-physical design and implementation of distributed event-triggered secondary control in islanded microgrids. IEEE Trans. Ind. Appl. **55**(6), 5631–5642 (2019)
120. Hu, S., Yuan, P., Yue, D., Dou, C., Cheng, Z., Zhang, Y.: Attack-resilient event-triggered controller design of DC microgrids under dos attacks. IEEE Trans. Circ. Syst. I: Regular Papers **67**(2), 699–710 (2020)
121. Lu, J., Zhang, X., Zhang, B., Hou, X., Wang, P.: Distributed dynamic event-triggered control for voltage restoration and current sharing in DC microgrids. IEEE Trans. Sustain. Energy **13**(1), 619–628 (2022)
122. Peng, J., Fan, B., Tu, Z., Zhang, W., Liu, W.: Distributed periodic event-triggered optimal control of DC microgrids based on virtual incremental cost. IEEE/CAA J. Automatica Sinica **9**(4), 624–634 (2022)
123. Fan, B., Peng, J., Yang, Q., Liu, W.: Distributed periodic event-triggered algorithm for current sharing and voltage regulation in DC microgrids. IEEE Trans. Smart Grid **11**(1), 577–589 (2020)
124. Chen, Y., Qi, D., Li, Z., Wang, Z., Yang, X., Zhang, J.: Distributed event-triggered control for frequency restoration in islanded microgrids with reduced trigger condition checking. CSEE J. Power Energy Syst. https://doi.org/10.17775/CSEEJPES.2020.05730
125. Zhao, G., Jin, L., Wang, Y.: Distributed event-triggered secondary control for islanded microgrids with disturbances: a hybrid systems approach. IEEE Trans. Power Syst. **38**(2), 1420–1431 (2023)
126. Qian, Y.-Y., Wan, Y., Lin, Z., Shamash, Y.A., Davoudi, A.: Distributed dynamic event-triggered control of power buffers in DC microgrids. IEEE Trans. Syst. Man Cybern. Syst. **52**(12), 7748–7759 (2022)
127. Xie, Y., Lin, Z.: Distributed event-triggered secondary voltage control for microgrids with time delay. IEEE Trans. Syst. Man Cybern. Syst. **49**(8), 1582–1591 (2019)
128. Ding, L., Han, Q.L., Zhang, X.M.: Distributed secondary control for active power sharing and frequency regulation in islanded microgrids using an event-triggered communication mechanism. IEEE Trans. Ind. Inf. **15**(7), 3910–3922 (2019)
129. Dou, C., Yue, D., Guerrero, J.M.: Multiagent system-based event-triggered hybrid controls for high-security hybrid energy generation systems. IEEE Trans. Ind. Inf. **13**(2), 584–594 (2017)

130. Guo, F., Wang, L., Wen, C., Zhang, D., Xu, Q.: Distributed voltage restoration and current sharing control in islanded DC microgrid systems without continuous communication. IEEE Trans. Ind. Electron. **67**(4), 3043–3053 (2019)
131. Lian, Z., Deng, C., Wen, C., Guo, F., Lin, P., Jiang, W.: Distributed event-triggered control for frequency restoration and active power allocation in microgrids with varying communication time delays. IEEE Trans. Ind. Electron. **68**(9), 8367–8378 (2020)
132. Zhai, M., Sun, Q., Wang, R., Wang, B., Liu, S., Zhang, H.: Fully distributed fault-tolerant event-triggered control of microgrids under directed graphs. IEEE Trans. Netw. Sci. Eng. **9**(5), 3570–3579 (2022)
133. Zhai, M., Sun, Q., Wang, B., Liu, Z., Zhang, H.: Cooperative fault-estimation-based event-triggered fault-tolerant voltage restoration in islanded AC microgrids. IEEE Trans. Autom. Sci. Eng. (2022)

Chapter 2
Cooperative Event-Triggered Fault-Tolerant Voltage Restoration in Islanded AC Microgrids

Abstract In this chapter, the problem of secondary voltage restoration in an islanded microgrid (MG) is considered, in which the actuator of the distributed generators (DGs) may coexist with partial loss of effectiveness (PLOE) fault and bias fault. For each DG, an adaptive event-triggered fault-tolerant (ETFT) control protocol is designed to compensate for the effects of actuator faults in the DGs, thereby restoring the voltage to the reference value. The dependent event triggering mechanism saves the processor's computational resources. A desirable feature of the protocol proposed in this paper is that the control protocol relies only on relative information between neighboring DG's, independent of global information about the network graph, fault boundaries, and network scale. It means that the protocol is implemented in a fully distributed framework. The protocol fully applies to the common ETFT consensus control problem of linear multiagent systems (MASs) with actuator faults. Furthermore, comprehensive theoretical arguments for consensus stability and analysis of Zeno behavior ensure the approach's feasibility. The simulation results verify the effectiveness of the algorithm.

Keywords Adaptive control · Fault-tolerant control · Event-triggered control · Multi-agent systems · Microgrids · Secondary voltage control

2.1 Introduction

In this chapter, we aim to study the secondary voltage restoration of islanded MGs based on ETFT control. The feedback linearization approach is used to convert an isolated MG's secondary control problem into a MAS consensus problem. A DG in an MG is regarded as an agent in a MAS. To restore the output voltage amplitude of each DG to the reference value, an adaptive ETFT secondary voltage control protocol is designed. The proposed adaptive control protocol requires only relative information between neighboring DGs. It is independent of the global information about the network graph, the global boundary of faults, and the scale of the network, which means that it is fully distributed and scalable. While effectively compensating for DGs' actuator faults, our approach updates the controller in an acyclic manner

J. Sun et al., *Building a Low-Carbon Future*, Studies in Systems, Decision and Control 506, https://doi.org/10.1007/978-3-031-43498-3_2

at the moment of discrete event sampling, reducing its update frequency and saving the consumption of processor computational resources. Utilizing boundary layer techniques in the control protocol avoids the undesirable chattering effect caused by the imperfect switching of devices by the nonsmooth static protocol.

2.2 Problem Formulation and Preliminaries

The information flow of DGs is described by a graph $\mathbb{G}$ which is composed of the node set $\mathbb{V} = \{\nu_0, \nu_1, \ldots, \nu_N\}$, the edge set $\mathcal{E} \subseteq \mathbb{V} \times \mathbb{V}$ and a weighted adjacency matrix $\mathcal{A} = [a_{ij}] \in \mathcal{R}^{(N+1)\times(N+1)}$ with non-negative adjacency elements a_{ij}. If and only if $(\nu_i, \nu_j) \in \mathcal{E}$, which means that node i can receive information from node j, then, $a_{ij} > 0$; otherwise, $a_{ij} = 0$. If the graph is undirected, $a_{ij} = a_{ji}$. Let the Laplacian matrix $\mathcal{L}$ be defined as $l_{ii} = \sum_{j=0}^{N} a_{ij}$ and $l_{ij} = -a_{ij}$, where $i \neq j$. If the root node has a path to every other node in the graph, then we can say that the graph has a spanning tree.

For secondary voltage control of MGs, define the voltage reference as a virtual leader and denote its communication with ith DG as a_{i0}.

Assumption 2.1 The subgraph associated with the followers is undirected and the graph $\mathbb{G}$ contains a directed spanning tree with the virtual leader as the root.

Then, $\mathcal{L}$ can be defined as the following structure:

$$\mathcal{L} = \begin{bmatrix} 0 & 0_{1\times N} \\ \mathcal{L}_2 & \mathcal{L}_1 \end{bmatrix}, \tag{2.1}$$

where $\mathcal{L}_2 \in \mathcal{R}^{N\times 1}$, $\mathcal{L}_1 \in \mathcal{R}^{N\times N}$. The real part of each eigenvalue of $\mathcal{L}_1$ is positive.

2.2.1 *System Model of an MG for Secondary Voltage Control*

Consider an islanded MG containing N DGs with a communication diagram of $\mathbb{G}$. For the ith DG system shown in Fig. 2.1, the dynamic model includes droop control, voltage, and current controllers, an LC filter, and line model.

The dynamics of the droop controller can be described as follows:

$$\begin{aligned} \omega_i &= \omega_{ni} - m_{P_i} P_i, \\ v_{odi}^* &= V_{ni} - n_{Q_i} Q_i, \\ v_{oqi}^* &= 0, \end{aligned} \tag{2.2}$$

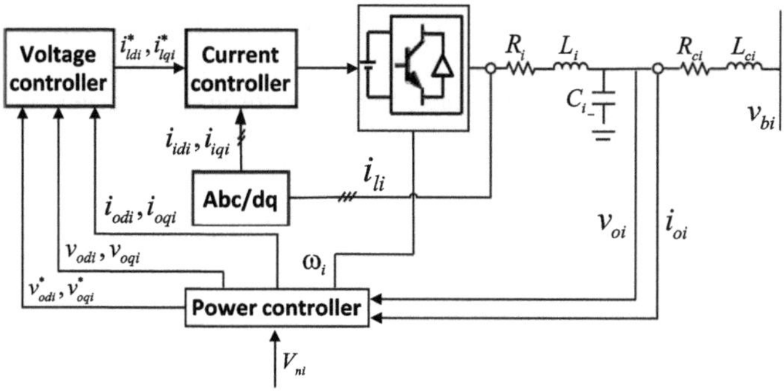

Fig. 2.1 Block diagram of an inverter-based DG

where ω_i represents the angular frequency, which is determined by the primary control. v^*_{odi} serves as the reference value for the DG's internal voltage control loop, determining the desired output voltage magnitude. ω_{ni} and V_{ni} correspond to the frequency and voltage inputs, respectively. The droop coefficients, m_{P_i} and n_{Q_i}, govern the sensitivity of the DG's power output to frequency and voltage variations. P_i and Q_i denote the active and reactive power at the terminals of the ith inverter.

The power controller dynamics can be described by the following equations:

$$\begin{aligned}\dot{P}_i &= \omega_{c_i}\left(v_{odi}i_{odi} + v_{oqi}i_{oqi}\right) - P_i\omega_{c_i},\\ \dot{Q}_i &= \omega_{c_i}\left(v_{oqi}i_{odi} - v_{odi}i_{oqi}\right) - Q_i\omega_{c_i},\end{aligned} \tag{2.3}$$

the parameter ω_{ci} represents the cutoff frequency of the low-pass filters utilized for power measurement. The variables v_{odi}, v_{oqi}, i_{odi}, and i_{oqi} correspond to the components of the output voltage v_{oi} and current i_{oi} of the ith inverter.

The voltage and current controller dynamics are

$$\begin{aligned}\dot{\phi}_{v_{di}} &= v^*_{odi} - v_{odi},\\ \dot{\phi}_{v_{qi}} &= v^*_{oqi} - v_{oqi},\\ \dot{\phi}_{i_{di}} &= i^*_{ldi} - i_{ldi},\\ \dot{\phi}_{i_{qi}} &= i^*_{lqi} - i_{lqi},\\ i^*_{ldi} &= K_{PVi}(v^*_{odi} - v_{odi}) + K_{IVi}\phi_{v_{di}} - \omega_i C_i v_{oqi} + F_{li}i_{odi},\\ i^*_{lqi} &= K_{PVi}(v^*_{oqi} - v_{oqi}) + K_{IVi}\phi_{v_{qi}} + \omega_i C_i v_{odi} + F_{li}i_{oqi},\\ v^*_{idi} &= K_{PCi}(i^*_{ldi} - i_{ldi}) + K_{ICi}\phi_{i_{di}} - \omega_i L_i i_{lqi},\\ v^*_{iqi} &= K_{PCi}(i^*_{lqi} - i_{lqi}) + K_{ICi}\phi_{i_{qi}} + \omega_i L_i i_{ldi}.\end{aligned} \tag{2.4}$$

where $\phi_{v_{di}}$, $\phi_{v_{qi}}$ are auxiliary state variables in the voltage controller. $\phi_{i_{di}}$, $\phi_{i_{qi}}$ are auxiliary state variables in the current controller. i^*_{ldi}, i^*_{lqi} are current reference in the current controller. i_{ldi} and i_{lqi} are filter current components. K_{PVi} and K_{IVi} are proportion and integral parameters in the voltage controller, respectively. $\phi_{v_{di}}, \phi_{v_{qi}}$ are auxiliary state variables in the voltage controller. F_{li} is feedforward parameter in the voltage controller. C_i and L_i are filter capacitor and inductance, respectively. K_{PCi} and K_{ICi} are proportion and integral parameters in the current controller, respectively. $\phi_{i_{di}}, \phi_{i_{qi}}$ are auxiliary state variables in the current controller.

The output LC filter and connector dynamics are

$$
\begin{aligned}
\dot{i}_{ldi} &= -\frac{R_i}{L_i} i_{ldi} + \omega_i i_{lqi} + \frac{1}{L_i} v_{idi} - \frac{1}{L_i} v_{odi} \\
\dot{i}_{lqi} &= -\frac{R_i}{L_i} i_{lqi} - \omega_i i_{ldi} - \frac{1}{L_i} v_{oqi} \\
\dot{v}_{odi} &= \omega_i v_{oqi} + \frac{1}{C_i} i_{ldi} - \frac{1}{C_i} i_{odi} \\
\dot{v}_{oqi} &= -\omega_i v_{odi} + \frac{1}{C_i} i_{lqi} - \frac{1}{C_i} i_{oqi} \\
\dot{i}_{odi} &= \frac{1}{L_{C_i}} v_{odi} - \frac{R_{C_i}}{L_{C_i}} i_{odi} - \frac{1}{L_{C_i}} v_{bdi} + \omega_i i_{oqi} \\
\dot{i}_{oqi} &= \frac{1}{L_{C_i}} v_{oqi} - \frac{R_{C_i}}{L_{C_i}} i_{oqi} - \frac{1}{L_{C_i}} v_{bqi} - \omega_i i_{odi}.
\end{aligned} \tag{2.5}
$$

where R_i is filter resistance. v_{idi}, v_{iqi} are output of current controller. v_{bdi}, v_{bqi} are the voltage at the PCC. L_{C_i}, R_{C_i} are inductance, resistance the line between the PCC and the inverter output.

The compact form of the nonlinear dynamic model of the AC MG can be expressed as follows:

$$
\begin{aligned}
\dot{x}_i &= f_i(x_i) + k_i(x_i) J_i + g_i(x_i) V_{ni}, \\
y_i &= d_i(x_i),
\end{aligned} \tag{2.6}
$$

where the state vector is

$$
x_i = [\delta_i, P_i, Q_i, \phi_{v_{di}}, \phi_{v_{qi}}, \phi_{i_{di}}, \phi_{i_{qi}}, i_{ldi}, i_{lqi}, v_{odi}, v_{oqi} i_{odi}, i_{oqi}]^T, \tag{2.7}
$$

and $J_i = [\omega_{com}, v_{bdi}, v_{bqi}]$. Detailed expressions for $f_i(x_i)$, $g_i(x_i)$, and $k_i(x_i)$ can be extracted from (2.2)–(2.5).

By utilizing feedback linearization, we obtain the following:

$$
\ddot{v}_{odi} = L_{F_i}^2 d_i + L_{g_i} L_{F_i} d_i V_{ni} = v_i, \tag{2.8}
$$

where $F_i = f_i(z_i) + k_i(z_i) J_i$, $L_{F_i} d_i = \frac{\partial d_i}{\partial x_i} F_i$ and $L_{F_i}^2 d_i = L_{F_i}(L_{F_i} d_i) = \frac{\partial L_{F_i} d_i}{\partial x_i} F_i$ are the Lie derivatives of d_i along F_i.

The control input V_{ni} is implemented by v_i as

$$V_{ni} = (L_{g_i} L_{F_i} d_i)^{-1}(-L_{F_i}^2 d_i + v_i). \tag{2.9}$$

Then, from (2.8), we have

$$\dot{z}_i(t) = Az_i(t) + Bv_i, i = 1, \ldots, N, \tag{2.10}$$

where $x_i = [v_{odi}, \dot{v}_{odi}]^T$, $A = \begin{bmatrix} 0 & 1 \\ 0 & 0 \end{bmatrix}$, $B = [0, 1]^T$, $C = [1, 0]$.

The virtual leader (reference value) dynamics of v_{odi} can be expressed as

$$\dot{z}_0(t) = Az_0(t), \tag{2.11}$$

where $z_0 = [y_0, y_{0,1}]^T = [v_{ref}, 0]^T$, and v_{ref} is the voltage reference.

The purpose of the secondary control is to design the control protocol of V_{ni}, so that the voltage v_{odi} can reach the reference value v_{ref}, that is, $z_i \to z_0$.

To realize the proposed control objective, a local neighborhood consensus error is defined, described as

$$\xi_i = \sum_{j=0}^{N} a_{ij} \left(z_i(t) - z_j(t) \right), i = 1, \ldots, N. \tag{2.12}$$

For the MG containing N DGs, the global disagreement vector ξ can be written as

$$\xi = (\mathcal{L}_1 \otimes I_n) \begin{bmatrix} z_1 - z_0 \\ \vdots \\ z_N - z_0 \end{bmatrix}, \tag{2.13}$$

where $\xi = \left[\xi_1^T(t), \xi_2^T(t), \ldots, \xi_N^T(t)\right]^T$.

Lemma 2.1 *[1] If ζ_1 and ζ_2 are positive constants, and $G(t)$ satisfies $\dot{G}(t) \leq -\zeta_1 G(t) + \zeta_2$, then*

$$G(t) \leq [G(0) - \frac{\zeta_2}{\zeta_1}]e^{-\zeta_1 t} + \frac{\zeta_2}{\zeta_1}. \tag{2.14}$$

2.3 Distributed Adaptive ETFT Protocol

2.3.1 Protocol Design Framework

Actuator faults of DGs occurring in MG are considered in this chapter. The actuator faults include PLOE fault and bias fault, which can be modeled as multiplicative fault and additive fault, as follows:

$$v_i^F = \theta_i v_i + v_i^S, i = 1, \ldots, N, \tag{2.15}$$

where θ_i, with $0 < \underline{\theta}_i \leq \theta_i \leq \overline{\theta}_i$, is the PLOE fault factor of the ith DG's actuator. $\underline{\theta}_i$ and $\overline{\theta}_i$ are known positive constants that stand for lower and upper bounds of θ_i. v_i^S is the bounded time-varying function referring to the bias fault factor of ith DG, which satisfies $\|v_i^S\| \leq \varsigma_i$ with ς_i being positive constants.

With consideration of the faults model, the dynamic of the ith DG is described as follows:

$$\dot{z}_i(t) = Az_i(t) + B(\theta_i v_i + v_i^S), i = 1, \ldots, N. \tag{2.16}$$

We define the following measurement error $e_i(t)$:

$$e_i(t) = \xi_i(t_k^i) - \xi_i(t), i = 1, \ldots, N, \tag{2.17}$$

where t_k^i denotes the kth triggering instant of ith DG. Then, the sequence of triggered instants is determined by the following trigger mechanism:

$$t_{k+1}^i = inf\{t > t_k^i : f_i(e_i, \xi_i(t_k^i)) > 0\}. \tag{2.18}$$

The triggering function of ith DG is designed as

$$\begin{aligned} f_i(t) = &\left[\tau(1 + \eta_1 + \eta_2)\phi_i(t)\bar{\theta}_i + \frac{\eta_3}{2}\underline{\theta}_i\right] \|B^T P\|^2 \|e_i(t)\|^2 \\ &+ 2\iota\left(\kappa_i \bar{\theta}_i + \varsigma\right) \|B^T P\| \|e_i(t)\| - \frac{1}{2}\underline{\theta}_i \xi_i^T(t_k^i) P B B^T P \xi_i(t_k^i) - o, \end{aligned} \tag{2.19}$$

where τ, o, ι are positive constants, $\kappa_i > \varsigma_i / \underline{\theta}_i$, $\eta_1, \eta_2, \eta_3 > 1$ and satisfy $\frac{1}{\eta_1} + \frac{1}{\eta_2} < 1$.

Based on the local information of neighbor DGs, a distributed adaptive control protocol is proposed for each DG. Let v_i in (2.15) be

$$v_i = -\phi_i(t) B^T P \xi_i(t_k^i) - \kappa_i f(B^T P \xi_i(t_k^i)), \tag{2.20}$$

where

$$f(\vartheta) = \begin{cases} \frac{\vartheta}{\|\vartheta\|}, & \|\vartheta\| > \chi_i \\ \frac{\vartheta}{\chi_i}, & \|\vartheta\| \leq \chi_i, \end{cases} \tag{2.21}$$

where χ_i is a small positive scalar, which is relevant to the width of the boundary layer.

The updating law of coupling time-varying consensus protocol gain ϕ_i in (2.20) is

$$\dot{\phi}_i = -k_{\phi_i}(\phi_i - 1) + \gamma_{\phi_i} \hat{\theta}_i(t) \xi_i^T(t_k^i) P B B^T P \xi_i(t_k^i), \tag{2.22}$$

where k_{ϕ_i}, γ_{ϕ_i} are positive constants, $\hat{\theta}_i$ is the estimation of unknown fault factor θ_i. Note that $\phi_i(0) \geq 1$, it can be easily obtained that $\phi_i(t) \geq 1$.

The parameter $\hat{\theta}_i$ in (2.22) is determined according to the adaptive law

$$\dot{\hat{\theta}}_i(t) = Proj_{[\underline{\theta}_i, \overline{\theta}_i]}\{\hat{\theta}_i\} = \begin{cases} 0, & if\ \hat{\theta}_i = \underline{\theta}_i\ and\ \hat{\theta}_{id} \leq 0 \\ & or\ \hat{\theta}_i = \overline{\theta}_i\ and\ \hat{\theta}_{id} \geq 0 \\ \gamma_{\theta_i}\hat{\theta}_{id}, & otherwise, \end{cases} \tag{2.23}$$

where the $Proj\{*\}$ represents the projection operator [2], whose purpose is to project the parameter $\hat{\theta}_i(t)$ to the interval $[\overline{\theta}_i, \underline{\theta}_i]$. γ_{θ_i} is a positive constant, and

$$\hat{\theta}_{id} = -\phi_i(t)\xi_i^T(t_k^i)PBB^TP\xi_i(t_k^i). \tag{2.24}$$

Then,

$$\dot{z}_i(t) = Az_i(t) - \theta_i\phi_i(t)BB^TP\xi_i(t_k^i) - \kappa_i\theta_iBf(B^TP\xi_i(t_k^i)) + Bv_i^S. \tag{2.25}$$

The closed-loop dynamics of the consensus error ξ can be expressed as:

$$\dot{\xi} = (I_N \otimes A)\xi - [\mathcal{L}_1\Theta\Phi(t) \otimes BB^TP]\tilde{\xi} - (\mathcal{L}_1\mathcal{K}\Theta \otimes B)F(\tilde{\delta}) + (\mathcal{L}_1 \otimes B)v^S, \tag{2.26}$$

where $\Theta = diag\{\theta_1, \theta_2, \cdots, \theta_N\}$, $\Phi(t) = diag\{\phi_1(t), \phi_2(t), \cdots, \phi_N(t)\}$, $\tilde{\xi} = [\xi_1^T(t_k^1), \xi_2^T(t_k^2), \cdots, \xi_N^T(t_k^N)]^T$, $\mathcal{K} = diag\{\kappa_1, \kappa_2, \cdots, \kappa_N\}$, $F(\tilde{\delta}) = col\{f(\tilde{\delta}_1)),$ $f(\tilde{\delta}_2), \cdots, f(\tilde{\delta}_N)\}$, $\tilde{\delta}_i = B^TP\xi_i(t_k^i)$, $v^S = col\{v_1^S, v_2^S, \cdots, v_N^S\}$.

2.3.2 *Consensus Analysis*

Theorem 2.1 *Suppose that Assumption 2.1 hold. Considering the adaptive closed-loop system described by (2.25), the ETFT controller (2.20) has the adaptive laws of (2.22) and (2.23). Then, all signals of the closed-loop system are uniformly constrained under the trigger condition (2.19), that is, the output voltages of all DGs can restore to the reference value v_{ref}. $P > 0$ in the control protocol (2.19) and (2.20) is a solution of the following algebraic Riccati equation:*

$$PA^T + AP - PBB^TP + I_N = 0. \tag{2.27}$$

Proof The following candidate Lyapunov functions are considered:

$$V = \xi^T(t)(\mathcal{L}_1^{-1} \otimes P)\xi(t) + \sum_{i=1}^{N}\frac{\tilde{\theta}_i^2}{2\gamma_{\theta_i}} + \sum_{i=1}^{N}\frac{(\phi_i(t) - \varpi)^2}{2\gamma_{\phi_i}},$$

where $\tilde{\theta}_i = \hat{\theta}_i - \theta_i$. Since the communication graph $\mathbb{G}$ satisfies Assumption 2.1, it is easy to obtain that $\mathcal{L}_1$ is non-singular. Then, V is positive definite.

Taking the time derivative of V along the trajectories of (2.26), one obtains:

$$\begin{aligned}\dot{V} &= 2\xi^T(\mathcal{L}_1^{-1} \otimes P)[(I_N \otimes A)\xi - (\mathcal{L}_1 \Theta \Phi(t) \otimes BB^T P)\tilde{\xi} + (\mathcal{L}_1 \otimes B)v^S \\ &\quad - (\mathcal{L}_1 \mathcal{K} \Theta \otimes B)F(\tilde{\delta})] + \sum_{i=1}^{N} \frac{\tilde{\theta}_i \dot{\hat{\theta}}_i}{\gamma_{\theta_i}} + \sum_{i=1}^{N} \frac{(\phi_i(t) - \varpi)\dot{\phi}_i}{\gamma_{\phi_i}} \\ &= \xi^T[\mathcal{L}_1^{-1} \otimes (PA + A^T P)]\xi - 2\xi^T(\Theta \Phi(t) \otimes PBB^T P)\tilde{\xi} \\ &\quad + 2\xi^T(I_N \otimes PB)v^S - 2\xi^T(\mathcal{K}\Theta \otimes PB)F(\tilde{\delta}) + \sum_{i=1}^{N} \frac{\tilde{\theta}_i \dot{\hat{\theta}}_i}{\gamma_{\theta_i}} \\ &\quad + \sum_{i=1}^{N} \frac{(\phi_i(t) - \varpi)\dot{\phi}_i}{\gamma_{\phi_i}}.\end{aligned} \tag{2.28}$$

Denote $e = \tilde{\xi} - \xi$, $e = [e_1^T(t), e_2^T(t), \ldots, e_N^T(t)]^T$. By Young's inequality, we have

$$\begin{aligned}&- 2\xi^T(\Theta \Phi(t) \otimes PBB^T P)\tilde{\xi} \\ &= -2\xi^T(\Theta \Phi(t) \otimes PBB^T P)\xi - 2\xi^T(\Theta \Phi(t) \otimes PBB^T P)e \\ &\leq -(2 - \frac{1}{\eta_1}) \sum_{i=1}^{N} \theta_i \phi_i \xi_i^T(t) PBB^T P \xi_i(t) + \eta_1 \sum_{i=1}^{N} \theta_i \phi_i \|B^T P\|^2 \|e_i\|^2.\end{aligned} \tag{2.29}$$

According to (2.20), one has the following:

$$\sum_{i=1}^{N} (\hat{\theta}_i - \theta_i)(-\hat{\theta}_{id}) + \sum_{i=1}^{N} \frac{(\hat{\theta}_i - \theta_i)\dot{\hat{\theta}}_i}{\gamma_{\theta_i}} \leq 0. \tag{2.30}$$

If $\hat{\theta}_i = \underline{\theta}_i$ and $\hat{\theta}_{id} \leq 0$ or $\hat{\theta}_i = \overline{\theta}_i$ and $\hat{\theta}_{id} \geq 0$, one can obtains that $\dot{\hat{\theta}}_i = 0$. Therefore, (2.30) is satisfied. Otherwise, $\dot{\hat{\theta}}_i = \gamma_{\theta_i} \hat{\theta}_{id}$, which implies that $\sum_{i=1}^{N}((\hat{\theta}_i - \theta_i)(-\hat{\theta}_{id}) + \sum_{i=1}^{N} \frac{(\hat{\theta}_i - \theta_i)\dot{\hat{\theta}}_i}{\gamma_{\theta_i}} = 0$.

$$
\begin{aligned}
&\sum_{i=1}^{N}\frac{\tilde{\theta}_i\dot{\tilde{\theta}}_i}{\gamma_{\theta_i}}+\sum_{i=1}^{N}\frac{(\phi_i(t)-\varpi)\dot{\phi}_i}{\gamma_{\phi_i}}\\
&\le-\sum_{i=1}^{N}(\hat{\theta}_i-\theta_i)\phi_i(t)\xi_i^T(t_k^i)PBB^TP\xi_i(t_k^i)\\
&\quad+\sum_{i=1}^{N}(\phi_i(t)-\varpi)\hat{\theta}_i\xi_i^T(t_k^i)PBB^TP\xi_i(t_k^i)-\sum_{i=1}^{N}\frac{k_{\phi_i}(\phi_i(t)-\varpi)(\phi_i-1)}{\gamma_{\phi_i}}\\
&\le\sum_{i=1}^{N}\theta_i\phi_i(t)\xi_i^T(t_k^i)PBB^TP\xi_i(t_k^i)-\sum_{i=1}^{N}\varpi\underline{\theta}_i\xi_i^T(t_k^i)PBB^TP\xi_i(t_k^i)\\
&\quad+\sum_{i=1}^{N}\frac{k_{\phi_i}}{\gamma_{\phi_i}}[-(\phi_i(t)-\varpi)^2-(\phi_i(t)-\varpi)(\varpi-1)]\\
&\le(1+\frac{1}{\eta_2})\sum_{i=1}^{N}\theta_i\phi_i(t)\xi_i^T(t)PBB^TP\xi_i(t)\\
&\quad+(1+\eta_2)\sum_{i=1}^{N}\theta_i\phi_i(t)\|B^TP\|^2\|e_i(t)\|^2\\
&\quad-\frac{1}{2}(1-\frac{1}{\eta_3})\sum_{i=1}^{N}\varpi\underline{\theta}_i\xi_i^T(t)PBB^TP\xi_i(t)+\frac{\eta_3}{2}\sum_{i=1}^{N}\varpi\underline{\theta}_i\|B^TP\|^2\|e_i(t)\|^2\\
&\quad-\frac{1}{2}\sum_{i=1}^{N}\varpi\underline{\theta}_i\xi_i^T(t_k^i)PBB^TP\xi_i(t_k^i)+\sum_{i=1}^{N}\frac{k_{\phi_i}}{2\gamma_{\phi_i}}[-(\phi_i(t)-\varpi)^2+(\varpi-1)^2].
\end{aligned}
\tag{2.31}
$$

According to the definition of θ_i and $f(\vartheta)$, we get

Case 1: $\|B^TP\xi_i(t_k^i)\|>\chi_i,\ i=1,2,\ldots,N$

$$
\begin{aligned}
&-2\xi^T(\mathcal{K}\Theta\otimes PB)F(\tilde{\delta})+2\xi^T(I_N\otimes PB)\upsilon^S\\
&\le-2\sum_{i=1}^{N}\kappa_i\theta_i\frac{\xi_i^T(t)PBB^TP\xi_i(t_k^i)}{\|B^TP\xi_i(t_k^i)\|}+2\sum_{i=1}^{N}\varsigma_i\|B^TP\xi_i(t)\|\\
&\le-2\sum_{i=1}^{N}\kappa_i\theta_i\frac{(\xi_i(t_k^i)-e_i)^TPBB^TP\xi_i(t_k^i)}{\|B^TP\xi_i(t_k^i)\|}+2\sum_{i=1}^{N}\varsigma_i\|B^TPe_i(t)\|\\
&\quad+2\sum_{i=1}^{N}\varsigma_i\|B^TP\xi_i(t_k^i)\|\\
&\le2\sum_{i=1}^{N}(\kappa_i\theta_i+\varsigma_i)\|B^TPe_i(t)\|-2\sum_{i=1}^{N}(\kappa_i\theta_i-\varsigma_i)\|B^TP\xi_i(t_k^i)\|.
\end{aligned}
\tag{2.32}
$$

Case 2: $\|B^T P\xi_i(t_k^i)\| \leq \chi_i, i = 1, 2, \ldots, N$

$$\begin{aligned}
&-2\xi^T(\mathcal{K}\Theta \otimes PB)F(\tilde{\delta}) + 2\xi^T(I_N \otimes PB)v^S \\
&\leq 2\sum_{i=1}^{N}\kappa_i\theta_i \frac{\|\xi_i^T(t)PB\|\|B^T P\xi_i(t_k^i)\|}{\chi_i} + 2\sum_{i=1}^{N}\varsigma_i\|B^T P\xi_i(t)\| \\
&\leq 2\sum_{i=1}^{N}(\kappa_i\theta_i + \varsigma_i)\|B^T P\xi_i(t)\| \\
&\leq 2\sum_{i=1}^{N}(\kappa_i\theta_i + \varsigma_i)\|B^T Pe_i(t)\| + 2\sum_{i=1}^{N}(\kappa_i\theta_i + \varsigma_i)\chi_i. \qquad (2.33)
\end{aligned}$$

Case 3: Without loss of generality, assume that $\|B^T P\xi_i(t_k^i)\| > \chi_i, i = 1, 2, \ldots, q$ and $\|B^T P\xi_i(t_k^i)\| \leq \chi_i, i = q+1, q+2, \ldots, N$, where $2 \leq q \leq N-1$, it can get

$$\begin{aligned}
&-2\xi^T(\mathcal{K}\Theta \otimes PB)F(\tilde{\delta}) + 2\xi^T(I_N \otimes PB)v^S \\
&\leq 2\sum_{i=1}^{N}(\kappa_i\theta_i + \varsigma_i)\|B^T Pe_i(t)\| - 2\sum_{i=1}^{q}(\kappa_i\theta_i - \varsigma_i)\|B^T P\xi_i(t_k^i)\| \\
&\quad + 2\sum_{i=q+1}^{N}(\kappa_i\theta_i + \varsigma_i)\chi_i. \qquad (2.34)
\end{aligned}$$

Choosing $\kappa_i > 0$ to be sufficiently large such that $\kappa_i > \varsigma_i/\underline{\theta}_i$ holds. From (2.32) (2.33) and (2.34), we can obtain

$$\begin{aligned}
&-2\xi^T(\mathcal{K}\Theta \otimes PB)F(\tilde{\delta}) + 2\xi^T(I_N \otimes PB)v^S \\
&\leq 2\sum_{i=1}^{N}(\kappa_i\theta_i + \varsigma_i)\|B^T P\|\|e_i(t)\| + 2\sum_{i=1}^{N}(\kappa_i\theta_i + \varsigma_i)\chi_i. \qquad (2.35)
\end{aligned}$$

In light of (2.29) (2.31) and (2.35), then, we can get that

$$\begin{aligned}\dot{V} \leq & \xi^T[\mathcal{L}_1^{-1} \otimes (PA + A^T P)]\xi \\ & - \sum_{i=1}^{N}[(1 - \frac{1}{\eta_1} - \frac{1}{\eta_2})\underline{\theta}_i\phi_i + \frac{1}{2}(1 - \frac{1}{\eta_3})\varpi\underline{\theta}_i]\xi_i^T(t)PBB^TP\xi_i(t) \\ & + \sum_{i=1}^{N}[(1 + \eta_2 + \eta_1)\bar{\theta}_i\phi_i + \frac{\eta_3}{2}\varpi\underline{\theta}_i]\|B^TP\|^2\|e_i\|^2 \\ & + 2\sum_{i=1}^{N}(\kappa_i\bar{\theta}_i + \varsigma_i)\|B^TP\|\|e_i(t)\| - \frac{1}{2}\sum_{i=1}^{N}\varpi\underline{\theta}_i\xi_i^T(t_k^i)PBB^TP\xi_i(t_k^i) \\ & + \sum_{i=1}^{N}\frac{k_{\phi_i}}{2\gamma_{\phi_i}}[-(\phi_i(t) - \varpi)^2 + (\varpi - 1)^2] + 2\sum_{i=1}^{N}(\kappa_i\bar{\theta}_i + \varsigma_i)\chi_i. \end{aligned} \tag{2.36}$$

It is easy to obtain that

$$\begin{aligned}\dot{V} \leq & \xi^T[\mathcal{L}_1^{-1} \otimes (PA + A^T P)]\xi \\ & - \sum_{i=1}^{N}[(1 - \frac{1}{\eta_1} - \frac{1}{\eta_2})\underline{\theta}_i\phi_i + \frac{1}{2}(1 - \frac{1}{\eta_3})\varpi\underline{\theta}_i]\xi_i^T(t)PBB^TP\xi_i(t) \\ & + \varpi\{\sum_{i=1}^{N}[\frac{1}{\tau\varpi}\tau(1 + \eta_1 + \eta_2)\phi_i(t)\bar{\theta}_i + \frac{\eta_3}{2}\underline{\theta}_i]\|B^TP\|^2\|e_i(t)\|^2 \\ & + 2\frac{1}{\iota\varpi}\iota(\kappa_i\bar{\theta}_i + \varsigma_i)\sum_{i=1}^{N}\|B^TP\|\|e_i(t)\| - \frac{1}{2}\sum_{i=1}^{N}\underline{\theta}_i\xi_i^T(t_k^i)PBB^TP\xi_i(t_k^i)\} \\ & + \sum_{i=1}^{N}\frac{k_{\phi_i}}{2\gamma_{\phi_i}}[-(\phi_i(t) - \varpi)^2 + (\varpi - 1)^2] + 2\sum_{i=1}^{N}(\kappa_i\bar{\theta}_i + \varsigma_i)\chi_i. \end{aligned} \tag{2.37}$$

Since $\phi_i(t) > 1$, choose $\frac{1}{\eta_1} + \frac{1}{\eta_2} < 1$ and $\varpi > 0$ to be sufficiently large such that $\varpi > \max\{2[\lambda_{min}^{-1}(\mathcal{L}_1)\min_{i=1,2,\cdots,N}\underline{\theta}_i^{-1} - (1 - \frac{1}{\eta_1} - \frac{1}{\eta_2})]/(1 - \frac{1}{\eta_3}), \frac{1}{\tau}, \frac{1}{\iota}\}$ holds, where $\lambda_{min}(\mathcal{L}_1)$ denotes the minimum eigenvalue of $\mathcal{L}_1$. Because of (2.27) and trigger function (2.19), it follows from (2.37) that

$$\begin{aligned}\dot{V} \leq & -\delta V - \xi^T[\mathcal{L}^{-1} \otimes (I_N - \delta P)]\xi + \varpi No + \delta\sum_{i=1}^{N}\frac{\tilde{\theta}_i^2}{2\gamma_{\theta_i}} \\ & + \sum_{i=1}^{N}(\delta - k_{\phi_i})\frac{(\phi_i(t) - \varpi)^2}{2\gamma_{\phi_i}} + \sum_{i=1}^{N}\frac{k_{\phi_i}}{2\gamma_{\phi_i}}(\varpi - 1)^2 + 2\sum_{i=1}^{N}(\kappa_i\bar{\theta}_i + \varsigma_i)\chi_i, \end{aligned} \tag{2.38}$$

where $0 < \delta \leq \min\{k_{\phi_i}, \lambda_{\max}^{-1}(P)\}$, such that $\dot{V}$ satisfies the following inequality:

$$\dot{V} \leq -\delta V + \varpi No + \delta \sum_{i=1}^{N} \frac{(\overline{\theta}_i - \underline{\theta}_i)^2}{2\gamma_{\theta_i}} + \sum_{i=1}^{N} \frac{k_{\phi_i}}{2\gamma_{\phi_i}} (\varpi - 1)^2 + 2 \sum_{i=1}^{N} (\kappa_i \bar{\theta}_i + \varsigma_i) \chi_i. \tag{2.39}$$

Using the Lemma 2.1, one can obtain

$$\begin{aligned} V(t) \leq e^{-\delta t} [V(0) - \frac{1}{\delta} (\delta \sum_{i=1}^{N} \frac{(\overline{\theta}_i - \underline{\theta}_i)^2}{2\gamma_{\theta_i}} + \sum_{i=1}^{N} \frac{k_{\phi_i}}{2\gamma_{\phi_i}} (\varpi - 1)^2 \\ + \varpi No + 2 \sum_{i=1}^{N} (\kappa_i \bar{\theta}_i + \varsigma_i) \chi_i)] \\ + \frac{1}{\delta} [\delta \sum_{i=1}^{N} \frac{(\overline{\theta}_i - \underline{\theta}_i)^2}{2\gamma_{\theta_i}} + \sum_{i=1}^{N} \frac{k_{\phi_i}}{2\gamma_{\phi_i}} (\varpi - 1)^2 + \varpi No \\ + 2 \sum_{i=1}^{N} (\kappa_i \bar{\theta}_i + \varsigma_i) \chi_i]. \end{aligned} \tag{2.40}$$

V exponentially converges to a bounded set

$$\mathcal{D} \triangleq \left\{ \xi, \hat{\theta}_i, \phi_i : V \leq \frac{1}{\delta} \ell \right\} \tag{2.41}$$

with a convergence rate faster than $e^{-\delta t}$, where

$$\ell = \sum_{i=1}^{N} \frac{k_{\phi_i}}{2\gamma_{\phi_i}} (\varpi - 1)^2 + 2 \sum_{i=1}^{N} (\kappa_i \bar{\theta}_i + \varsigma_i) \chi_i + \varpi No. \tag{2.42}$$

Therefore ξ, ϕ_i and $\tilde{\theta}_i$, $i = 1, 2, \ldots, N$ are uniformly ultimately bounded and ξ exponentially converges to the residual set $\mathcal{D}$. The proof is complete. ■

Remark 2.1 The global nonzero eigenvalue information of the Laplacian matrix associated with the communication graph was generally required to determine some parameters in the control laws and/or the triggering functions. The adaptive protocol in this chapter is independent of any global information about the network graph or fault boundaries, as well as the network scale, which means that it is fully distributed and scalable.

Remark 2.2 Due to the use of the $f(\vartheta)$ function, the proposed protocol (2.20) avoids the undesirable chattering effect encountered by discontinuous controllers in [3, 4]. However, the cost is that the proposed continuous agreement will lead to convergence to a larger bounded set, because the residual set is associated with χ_i. Therefore, by (2.41), in order to ensure that the consensus error ξ converges progressively to a satisfactory vicinity of zero, we can choose a relatively small χ_i, k_{ϕ_i} and a relatively large γ_{θ_i}, γ_{ϕ_i}.

2.3.3 Zeno Phenomenon Exclusion

Theorem 2.2 *For system (2.25), Assumption 2.1 holds, and the control protocol consists of a controller (2.20) and a trigger rule (2.19), which can exclude the Zeno phenomenon.*

Proof Taking the time derivative on $\|e_i(t)\|$, one has:

$$D^+\|e_i(t)\| = \frac{d}{dt}\sqrt{e_i^T(t)e_i(t)} \leq \|\dot{e}_i(t)\|. \tag{2.43}$$

From the definition of $e_i(t)$, for $t \in [t_k^i, t_{k+1}^i)$, from (2.26), it is easily to obtain

$$\dot{e}_i(t) = -\dot{\xi}_i(t) = Ae_i(t) + \sigma_i. \tag{2.44}$$

where

$$\begin{aligned}\sigma_i =& - A\xi_i(t_k^i) - \sum_{j=1}^{N} l_{ij}\theta_j\phi_j(t)BB^T P\xi_j(t_k^j) \\ &- \sum_{j=1}^{N} l_{ij}\kappa_j\theta_j Bf(B^T P\xi_j(t_k^j)) + \sum_{j=1}^{N} l_{ij}Bv_j^S.\end{aligned} \tag{2.45}$$

It follows from Theorem 2.1 that both $\xi_i(t)$ and $\phi_i(t)$ are bounded. It's easy to see that σ_i is also bounded. Subsequently, denote $\bar{\sigma}_i$ as the maximum value of the $\|\sigma_i\|$, then

$$\frac{d\|e_i(t)\|}{dt} \leq \|A\|\|e_i(t)\| + \bar{\sigma}_i. \tag{2.46}$$

Note that, $e_i(t_k^i) = 0$. By Comparison Principle, for $t \in [t_k^i, t_{k+1}^i]$, $\|e_i(t)\|$ satisfies the following inequality:

$$\|e_i(t)\| = (\bar{\sigma}_i/\|A\|)(e^{\|A\|(t-t_k^i)} - 1). \tag{2.47}$$

Trigger function (2.19) satisfies $f_i(t) \leq 0$, if the following conditions is established:

$$\|e_i(t)\| \leq \frac{-\left[2\iota(\kappa_i\bar{\theta}_i + \varsigma)\right] + \sqrt{\Delta}}{2\left[\tau(1+\eta_1+\eta_2)\phi_i(t)\bar{\theta}_i + \frac{\eta_3}{2}\underline{\theta}_i\right]\|B^T P\|} \triangleq e_i^*, \tag{2.48}$$

where

$$\Delta = \left[2\iota(\kappa_i\bar{\theta}_i + \varsigma)\right]^2 + 4\left[\tau(1+\eta_1+\eta_2)\phi_i(t)\bar{\theta}_i + \frac{\eta_3}{2}\underline{\theta}_i\right] \times \left[\frac{1}{2}\underline{\theta}_i\xi_i^T(t_k^i)PBB^T P\xi_i(t_k^i) + o\right]. \tag{2.49}$$

Then, solve the following inequality:

$$\frac{\bar{\sigma}_i}{\|A\|}(e^{\|A\|\tau_k^i} - 1) \geq e_i^*, \tag{2.50}$$

a lower bound τ_k^i of $t_{k+1}^i - t_k^i$ can be obtained.

Based on (2.48) (2.49) and (2.50), it yields that

$$t_{k+1}^i - t_k^i \geq \tau_k^i \geq \frac{1}{\|A\|} ln(1 + \frac{\|A\|}{\bar{\sigma}_i} e_i^*). \tag{2.51}$$

Note that τ_k^i always exists and is strictly positive. Then, the Zeno phenomenon has been excluded. This completes the proof.

■

Remark 2.3 By (42)-(45), we derive the lower bound $\tau_k^i \geq \frac{1}{\|A\|}\ln(1 + \frac{\|A\|}{\bar{\sigma}_i}e_i^*)$ for $t_{k+1}^i - t_k^i$. If the exponential decay function is introduced in the event-triggered mechanism, it is easy to derive $\lim_{t\to\infty}\frac{1}{\|A\|}\ln(1 + \frac{\|A\|}{\bar{\sigma}_i}e_i^*) = 0$. This also means that the trigger interval can be arbitrarily small. However, the algorithm will fail when the event interval is more significant than zero and less than the sampling frequency of the device, which is a common disadvantage of the distributed event-based algorithms. Therefore, add a positive constant o to the trigger function so that $\tau_k^i \geq \frac{1}{\|A\|}\ln(1 + \frac{\|A\|}{\bar{\sigma}_i}e_i^*)$ is greater than a positive number. Thereby, the proposed ETFT protocol provides a positive minimum inter-event time at the cost of convergence to a more extensive bounded set.

As a result, the proposed control strategy ensures that the output voltages of all DGs are restored to the reference value.

2.4 Simulation Results

In this section, to demonstrate the effectiveness of the proposed ETFT secondary control protocol, various scenarios are simulated for an isolated MG in a MATLAB/Simulink software environment. The minimal communication swarm is shown

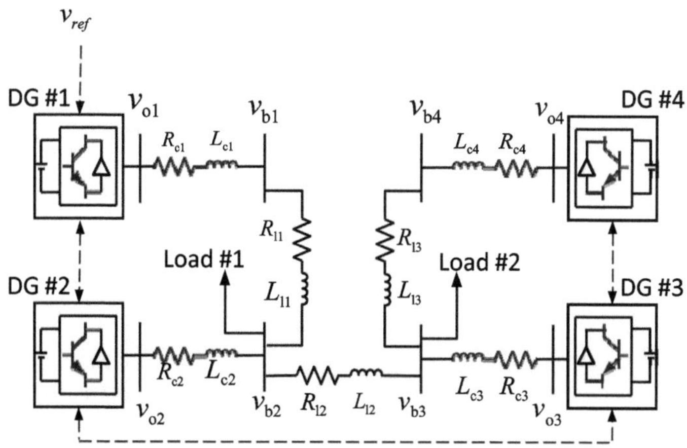

Fig. 2.2 The example MG test system and considered communication structure

in Fig. 2.2. The islanded MG test is a combination of four inverter-based DGs. The lines are modeled as series RL branches. Table 2.1 provides the DGs, lines, and loads parameters. In this table, K_{PV}, K_{IV}, K_{PC} and K_{IC} are the voltage and current controllers parameters. The simulation results are extracted by modeling the dynamical equations of MG in Matlab. Assume that the DGs communicate with each other via the undirected graph shown in Fig. 2.2. Only DG#1 communicates with the virtual leader (reference value), and it is a one-way communication.

For this communication network

$$\mathcal{L} = \begin{bmatrix} 0 & 0 & 0 & 0 & 0 \\ -1 & 2 & -1 & 0 & 0 \\ 0 & -1 & 2 & -1 & 0 \\ 0 & 0 & -1 & 2 & -1 \\ 0 & 0 & 0 & -1 & 1 \end{bmatrix}.$$

The control parameters are selected as

$$P = \begin{bmatrix} 1.7321 & 1.0000 \\ 1.0000 & 1.7321 \end{bmatrix},$$

and others are chosen as: $\tau = 2.2$, $\iota = 1.4$, $\kappa_1 = \kappa_2 = 1.3$, $\kappa_3 = 2.4$, $\kappa_3 = 3.1$, $\eta_1 = 3$, $\eta_2 = 2$, $\eta_3 = 2.5$, $\varsigma = 0.9$, $o = 10$, $\bar{\theta}_1 = 1.2$, $\bar{\theta}_2 = 1.3$, $\bar{\theta}_3 = \bar{\theta}_4 = 0.7$, $k_{\phi_i} = 1$, $\gamma_{\phi i} = \gamma_{\theta_i} = 10$, $i = 1, \ldots, 4$.

The analysis of Case 1 demonstrates the effectiveness of the proposed secondary control for voltage restoration to the reference value. The analysis of the case 2 demonstrates the fault tolerance and effectiveness of the proposed secondary control in the event of faults, moreover, verifies its PIPO capability.

Table 2.1 Specifications of the MG Test System

DGs	DG#1 & DG#2 & DG#3 & DG#4	
	m_P	1.5×10^{-5}
	n_Q	2×10^{-4}
	R_c	0.35Ω
	L_c	1.847 mH
	K_{PV}	0.2013
	K_{IV}	480
	K_{PC}	6.72
	K_{IC}	3360
Lines	$R_{l1} = R_{l2} = 0.35\Omega$	$R_{l1} = 0.23\Omega$
	$L_{l1} = L_{l2} = 1.847mH$	$L_{l1} = 0.318mH$
RL loads	Load #1 & Load #2	
	P(per phase)	Q(per phase)
	50 kW	12 kVAr

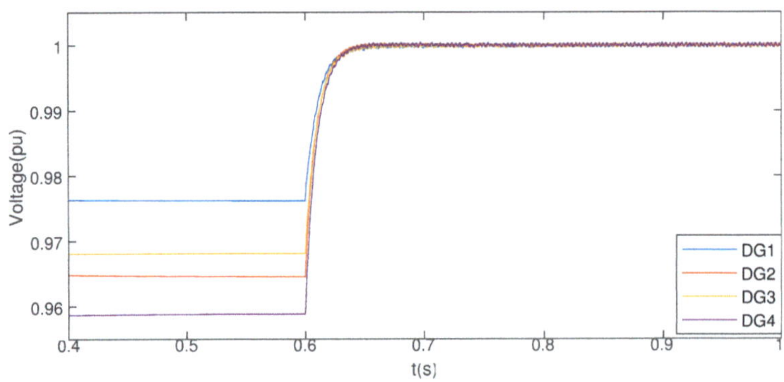

Fig. 2.3 $v_{ref} = 1p.u.$ in case 1: DG output voltage magnitudes

2.4.1 *Case 1: Effectiveness Evaluation for the ETFT Control Protocol*

To show the tracking performance of the proposed control method, the secondary control is intentionally deactivated at the beginning and only the primary control is applied. At $t = 0.6$ s, the proposed secondary control is activated.

The simulation results for the reference value $v_{ref} = 1p.u.$ is shown in Fig. 2.3. The output voltages of the four DGs deviate somewhat from the nominal values at the beginning of the simulation. Afterwards, the secondary controls are enabled at $t = 0.6$s, all terminal voltage amplitudes return to the reference value within 0.1 s.

The triggering instants for each DG is shown in Fig. 2.4, which shows no Zeno phenomenon occurs.

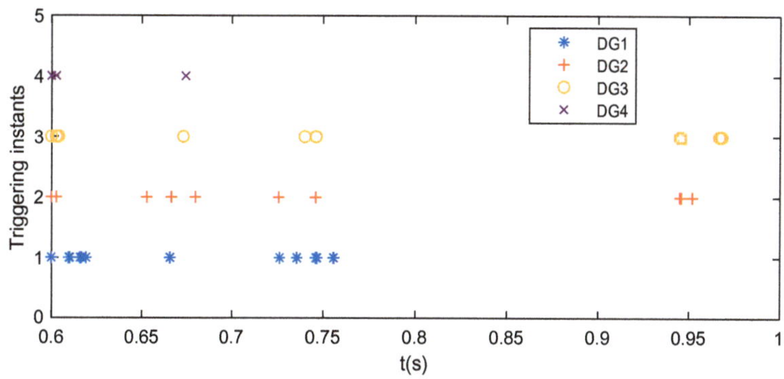

Fig. 2.4 $v_{ref} = 1p.u.$ in case 1: triggering instants for DGs

Therefore, the primary control can keep the voltage stable, but will cause a certain voltage deviation. The proposed ETFT secondary voltage control protocol can not only restore the voltage to the reference value, but also reduce the energy in communication.

2.4.2 Case 2: Fault Tolerance Evaluation and PIPO Capability for the ETFT Control Protocol

The PLOE faults and bias faults in the DGs are chosen as follows: $\theta_1 = 1, \theta_2 = 1.2$, $\theta_3 = 0.44$ and $\theta_4 = 0.64$, $v_1^s = 0.2\cos(2t - 3) + 0.2$, $v_2^s = 0.3\sin(2t - 3) + 0.1$, $v_3^s = 0.1\cos(2t - 10) + 0.4$, $v_4^s = 0.4\cos(3t - 2) + 0.1$.

Similarly, at $t = 0.6$ s, the secondary control is activated. Differently, at $t = 1.2$ s, DG#4 is withdrawn from operation and reconnected at $t = 2s$.

Figure 2.5 reveals that the proposed control performs well in fault tolerance for PLOE faults and bias faults. Moreover, the proposed control protocol scheme has good responsiveness and robustness to the disconnection and reconnection of the DG units. Although there are some transients after DG#4 disconnection and reconnection, it achieves the PIPO function of MG. The trigger moments for each DG are shown in Fig. 2.6, which shows no Zeno phenomenon occurs.

2.5 Conclusions

The feedback linearization method is applied to transform the secondary control problem of an isolated MG system into a tracking consensus problem of a MAS. That is, a DG in an MG system is considered an agent in a MAS. Then, an adaptive ETFT control protocol for the secondary voltage control of MGs is proposed, which

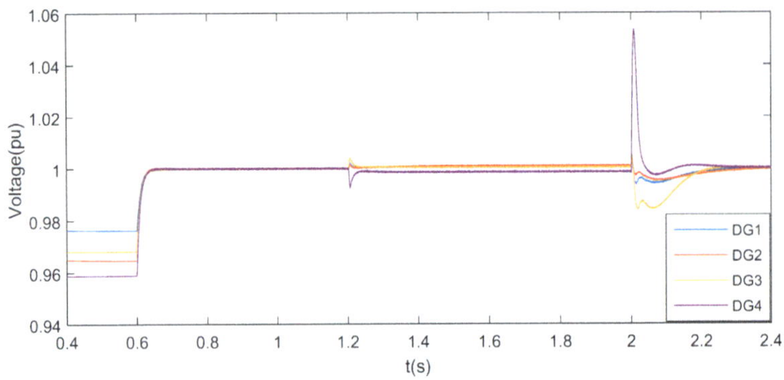

Fig. 2.5 $v_{ref} = 1p.u.$ in case 2: DG output voltage magnitudes

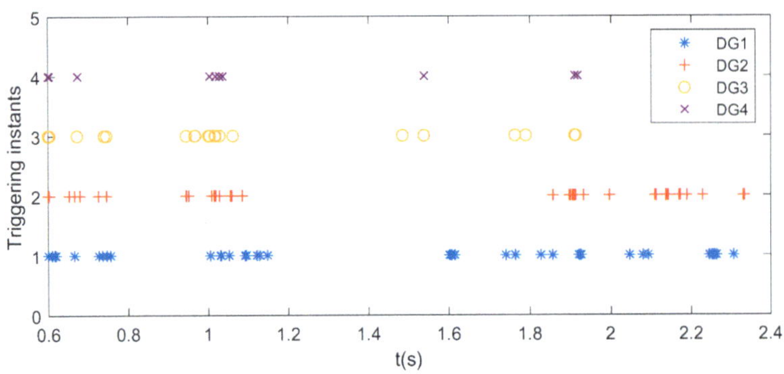

Fig. 2.6 $v_{ref} = 1p.u.$ in case 2: triggering instants for DGs

can restore the output voltage amplitude of all DGs to a reference value, and no Zeno behavior is exhibited. The method saves the consumption of processor computational resources while effectively suppressing the effects caused by DG actuator faults. In addition, it is worth noting that the proposed adaptive ETFT protocol is fully distributed and scalable. A Matlab/Simulink simulation system verifies the effectiveness of the proposed secondary control strategy.

References

1. Slotine, J., Li, W.: Applied nonlinear control. Prentice-Hall, Englewood Cliffs, NJ (1991)
2. Ioannou, P.A., Sun, J.: Robust adaptive control. Prentice-Hall, Upper Saddle River, NJ, USA (2012)
3. Khalili, M., Zhang, X., Polycarpou, M.M., Parisini, T., Cao, Y.: Distributed adaptive fault-tolerant control of uncertain multi-agent systems. Automatica **87**, 142–151 (2018)
4. Luo, S., Ye, D.: Adaptive double event-triggered control for linear multi-agent systems with actuator faults. IEEE Transactions on Circuits and Systems I: Regular Papers **66**(12), 4829–4839 (2019)

Chapter 3
Fully Distributed Fault-Tolerant Event-Triggered Control of Microgrids Under Directed Graphs

Abstract This chapter proposes an event-triggered fault tolerant (ETFT) control strategy for voltage restoration in microgrids (MGs). The presence of partial loss of effectiveness (PLOE) and bias faults in the actuators are considered. The proposed algorithm is independent of the global information related to the directed communication network and does not involve the boundaries of the fault factors. Each distributed generator (DG) only needs to know the relative information between its neighbors, which means that the algorithm is fully distributed. The boundary layer technique is used to construct smooth ETFT controllers, avoiding the undesirable chattering effect of discontinuous controllers. Furthermore, a novel consensus error prediction model is developed to eliminate the need for continuous communication between DGs. The proposed algorithm is not only applicable to the secondary voltage restoration of MGs but can also be extended to general linear multi-agent systems (MASs). Finally, the simulation results verify the effectiveness of the algorithm.

Keywords Event-triggered control · Fault-tolerant control · Adaptive control · Multi-agent systems · Microgrids · Secondary voltage control

3.1 Introduction

In this chapter, we aim to study the secondary voltage restoration of islanded MGs based on ETFT control. The feedback linearization approach is used to convert the secondary control problem of an isolated MG into a consensus problem of a MAS. To restore the output voltage amplitude of all DGs to the reference value, for each DG, an adaptive ETFT secondary voltage control strategy is designed. A fully distributed ETFT consensus algorithm is proposed, based only on the relative information between neighboring DGs. The algorithm avoids the Laplacian and the scale of the network graphs. Moreover, the scheme proposed in this chapter can also be applied to general directed graphs containing a spanning tree. A smooth ETFT controller design process is proposed. Due to the introduction of adaptive and boundary layer techniques, the control algorithm no longer needs to know the boundaries of the fault factors, while avoiding the undesirable chatting effects encountered

J. Sun et al., *Building a Low-Carbon Future,* Studies in Systems, Decision and Control 506, https://doi.org/10.1007/978-3-031-43498-3_3

by non-smooth controllers. For general directed graphs of MASs, fully distributed event-triggered algorithms typically require two or more event-triggering conditions. Compared with a multi-event monitoring system, the event-triggered mechanism proposed in this chapter reduces the load of processing data. A consensus error prediction model to avoid continuous communication between DGs is proposed under the premise that the fault factors of DGs are known or invariable.

3.2 Modeling Framework

Consider an islanded MG containing N DGs with a communication diagram of $\mathcal{G}$. For the ith DG system shown in Fig. 2.1, the dynamic model includes droop control, inner-loop voltage and current control, an LC filter, and line model.

We model the information flow structures among DGs as a graph $\mathcal{G}$ which is composed of the node set $\mathbb{V} = \{\nu_0, \nu_1, \ldots, \nu_N\}$, the edge set $\mathcal{E} \subseteq \mathbb{V} \times \mathbb{V}$ and a weighted adjacency matrix $\mathcal{A} = [a_{ij}] \in \mathcal{R}^{(N+1)\times(N+1)}$ with non-negative adjacency elements a_{ij}. If and only if $(\nu_i, \nu_j) \in \mathcal{E}$, which means that node i can receive information from node j, then, $a_{ij} > 0$; otherwise, $a_{ij} = 0$. Let the Laplacian matrix $\mathcal{L}$ be defined as $\mathcal{L}_{ii} = \sum_{j=0}^{N} a_{ij}$ and $\mathcal{L}_{ij} = -a_{ij}$, where $i \neq j$. Thus, the Laplace matrix of the directed graph is asymmetric. If the root node has a path to every other node in the graph, then we can say that the graph has a spanning tree.

Referring to the well-known dynamics of MG containing N DGs in [1]. By feedback linearization, we can obtain that:

$$\ddot{v}_{odi} = f_i(x_i) + g_i(x_i)V_{ni}, i = 1, \cdots, N, \tag{3.1}$$

where v_{odi}, V_{ni} are the component of the output voltage amplitude v_{oi} and the voltage input in Fig. 2.1, $f_i(x_i)$ and $g_i(x_i)$ are meticulously described by Bidram et al. [1]. We define $u_i = f_i(x_i) + g_i(x_i)V_{ni}$. We discuss actuators with PLOE and bias faults that occur in MGs. Then, we have

$$\ddot{v}_{odi} = \mathcal{Q}_i u_i + \mathcal{P}_i, i = 1, \ldots, N, \tag{3.2}$$

where $\mathcal{Q}_i$, with $0 < \mathcal{Q}_i \leq 1$, is a PLOE fault factor of the ith DG's actuator. $\mathcal{P}_i$ is time-varying partial loss of effectiveness fault severity of the ith DG satisfying $\|\mathcal{P}_i\| \leq \mathcal{P}_i^S$ with $\mathcal{P}_i^S > 0$ being a constant.

Then, the dynamic model of the ith DG can be represented as follows:

$$\dot{x}_i(t) = Ax_i(t) + B(\mathcal{Q}_i u_i + \mathcal{P}_i), i = 1, \ldots, N. \tag{3.3}$$

where $x_i = [v_{odi}, \dot{v}_{odi}]^T$, $A = \begin{bmatrix} 0 & 1 \\ 0 & 0 \end{bmatrix}$, $B = [0, 1]^T$.

The virtual leader (reference) dynamics of v_{odi} can be expressed as

$$\dot{x}_0(t) = Ax_0(t), \tag{3.4}$$

where $x_0 = [v_{ref}, 0]^T$, v_{ref} is the voltage reference.

The purpose of the secondary control is to design the control algorithms of the virtual controller u_i, which is equivalent to designing the control algorithm of V_{ni}. Thus enabling the voltage v_{odi} to reach the reference value v_{ref}, that is, $y_i \to y_0$.

To realize the proposed control objective, a local neighborhood consensus error is described as

$$\vartheta_i = \sum_{j=0}^{N} a_{ij}(x_i(t) - x_j(t)), i = 1, \ldots, N. \tag{3.5}$$

Assumption 3.1 The communication graph $\mathcal{G}$ contains a directed spanning tree with the virtual leader as the root node.

Then, $\mathcal{L}$ can be defined as the following structure:

$$\mathcal{L} = \begin{bmatrix} 0 & 0_{1\times N} \\ \mathcal{L}_2 & \mathcal{L}_1 \end{bmatrix}, \tag{3.6}$$

where $\mathcal{L}_2 \in \mathcal{R}^{N\times 1}$, $\mathcal{L}_1 \in \mathcal{R}^{N\times N}$.

For the MG containing N DGs, the global disagreement vector ϑ of the graph $\mathcal{G}$ can be written as

$$\vartheta = (\mathcal{L}_1 \otimes I_n) \begin{bmatrix} x_1 - x_0 \\ \vdots \\ x_N - x_0 \end{bmatrix}, \tag{3.7}$$

where $\vartheta = [\vartheta_1^T(t), \vartheta_2^T(t), \ldots, \vartheta_N^T(t)]^T$.

Lemma 3.1 *[2] Consider a nonsingular M-matrix $\mathcal{L}_1$. There exists a diagonal matrix Ω so that $\Omega \equiv diag\{\omega_1, \omega_2, \ldots, \omega_N\}$ and $\Omega\mathcal{L}_1 + \mathcal{L}_1^T\Omega > 0$.*

Lemma 3.2 *[3] Let A and B be $m \times n$ and $n \times m$ complex matrices, respectively. Then AB and BA have the same nonzero eigenvalues.*

3.3 Adaptive ETFT Voltage Restoration Control

The purpose of this section is to propose a fully distributed ETFT control algorithm. Thereby, the voltage restoration does not depend on the global information of the directed communication network and the boundaries of the fault factors.

Denote the measurement error of $\vartheta_i(t)$ relative to the latest triggering instants t_k^i by $\zeta_i(t)$:

$$\zeta_i(t) = \vartheta_i(t_k^i) - \vartheta_i(t), i = 1, \ldots, N, \tag{3.8}$$

where t_k^i with $t_0^i = 0$, $k \in \mathbb{N}$, is the kth sampling instant of the controller i, For the ith agent, $t_0^i, t_1^i, \ldots, t_N^i$ is the event-triggered time sequence and determined by the following triggering condition:

$$t_{k+1}^i = inf\{t > t_k^i : f_i(t) > 0\}. \tag{3.9}$$

The triggering function for each DG is given by

$$\begin{aligned} f_i(t) =& \mathcal{M}_{1i}\|P\zeta_i(t)\|^2 + 2\beta_i(c+1)(\hat{\kappa}_i + \ell_i)\|B^T\|\|P\zeta_i(t)\| \\ & - \frac{1}{2}\vartheta_i^T(t_k^i)PBB^TP\vartheta_i(t_k^i) - \rho_i e^{-o_i t}. \end{aligned} \tag{3.10}$$

where

$$\begin{aligned} \mathcal{M}_{1i} =& \beta_i\mathcal{J}_i + [\beta_i(1+\varpi_3)(\hat{\kappa}_i + \ell_i) + \beta_i(\hat{\kappa}_i + \ell_i)^2 \\ & + \frac{2}{\chi_i}\beta_i(c+1)(1+\varpi_5^{-1})(\hat{\kappa}_i + \ell_i) + \frac{1}{2}\varpi_4^{-1}]\|B^T\|^2, \end{aligned} \tag{3.11}$$

with $\mathcal{J}_i = (\vartheta_i(t_k^i) + \vartheta_i(t))^T(\vartheta_i(t_k^i) + \vartheta_i(t))\vartheta_i^T(t_k^i)PBB^TP\vartheta_i(t_k^i)$. β_i, ϖ_3, ϖ_4 and ϖ_5 are positive constants. c is a positive constant. χ_i is designed later. $P > 0$ in the control algorithm is a solution of the following algebraic Riccati equation (ARE):

$$PA^T + AP - PBB^TP + I_N = 0. \tag{3.12}$$

Moreover,

$$\dot{\hat{\kappa}}_i = -\nu_{\kappa_i}(\hat{\kappa}_i - 1) + \vartheta_i^T(t_k^i)PBB^TP\vartheta_i(t_k^i), \tag{3.13}$$

$$\ell_i = \vartheta_i^T(t)P\vartheta_i(t). \tag{3.14}$$

Let the auxiliary control u_i in (3.3) be

$$u_i = -(\hat{\kappa}_i + \bar{\ell}_i)B^TP\vartheta_i(t_k^i) - c\eta(B^TP\vartheta_i(t_k^i)), \tag{3.15}$$

$$\bar{\ell}_i = \vartheta_i^T(t_k^i)P\vartheta_i(t_k^i), \tag{3.16}$$

$$\eta(*)=\begin{cases}\frac{*}{\|*\|}, & if\ \|*\|>\chi_i\\ \frac{\star}{\chi_i}, & if\ \|*\|\leq\chi_i,\end{cases} \tag{3.17}$$

where χ_i is a small positive scalar, which is relevant to the width of the boundary layer. Then,

$$\begin{aligned}\dot{x}_i(t)=&Ax_i(t)-\mathcal{Q}_i(\hat{\kappa}_i+\bar{\ell}_i)BB^TP\vartheta_i(t_k^i)\\&-c\mathcal{Q}_iB\eta(B^TP\vartheta_i(t_k^i))+B\mathcal{P}_i, i=1,\ldots,N.\end{aligned} \tag{3.18}$$

It is not difficult to obtain that ϑ satisfies the following dynamics:

$$\begin{aligned}\dot{\vartheta}=&(I_N\otimes A)\vartheta-\mathcal{L}_1\mathcal{Q}(\hat{\kappa}+\bar{\ell})\otimes BB^TP\tilde{\vartheta}\\&-c(\mathcal{L}_1\mathcal{Q}\otimes B)\eta(\tilde{\sigma})+(\mathcal{L}_1\otimes B)\mathcal{P},\end{aligned} \tag{3.19}$$

where $\vartheta=diag\{\vartheta_1,\vartheta_2,\ldots,\vartheta_N\}$, $\mathcal{Q}=diag\{\mathcal{Q}_1,\mathcal{Q}_2,\ldots,\mathcal{Q}_N\}$, $\hat{\kappa}=diag\{\hat{\kappa}_1,\hat{\kappa}_2,\ldots,\hat{\kappa}_N\}$, $\bar{\ell}=diag\{\bar{\ell}_1,\bar{\ell}_2,\ldots,\bar{\ell}_N\}$, $\tilde{\vartheta}=diag\{\vartheta_1(t_k^1),\vartheta_2(t_k^2),\ldots,\vartheta_N(t_k^N)\}$, $\eta(\tilde{\sigma})=[\eta(B^TP\vartheta_1(t_k^1)),\eta(B^TP\vartheta_2(t_k^2)),\ldots,\eta(B^TP\vartheta_N(t_k^N))]^T$ and $\mathcal{P}=diag\{\mathcal{P}_1,\mathcal{P}_2,\ldots,\mathcal{P}_N\}$.

Theorem 3.1 *Suppose that Assumption 3.1 hold. Considering the adaptive closed-loop system described by (3.3), the ETFT controller (3.15) has the adaptive laws of (3.13). Then, all signals of the closed-loop system are uniformly constrained under the trigger condition (3.10), that is, the output voltages of all DGs can restore to the reference value* v_{ref}.

Proof Consider the following Lyapunov function candidate

$$V=\frac{1}{2}\sum_{i=1}^{N}\omega_i\mathcal{Q}_i(2\hat{\kappa}_i+\ell_i)\ell_i+\frac{1}{2}\sum_{i=1}^{N}\omega_i\mathcal{Q}_i\tilde{\kappa}_i^2, \tag{3.20}$$

where $\tilde{\kappa}_i=\hat{\kappa}_i-\bar{\kappa}_i$, and $\bar{\kappa}_i$ is a positive constant to be determined later. Note that $\hat{\kappa}_i(0)\geq 1$ and $\dot{\hat{\kappa}}_i\geq 0$ when $\hat{\kappa}_i=1$ in (3.13). Then, it is not difficult to see that $\hat{\kappa}_i(t)\geq 1$ for any $t>0$. It is easy to see that V is the positive definite.

The time derivate of V along the trajectory of (3.18) is obtained as follows:

$$\begin{aligned}\dot{V}=&\sum_{i=1}^{N}\omega_i\mathcal{Q}_i(\hat{\kappa}_i+\ell_i)\dot{\ell}_i+\sum_{i=1}^{N}\omega_i\mathcal{Q}_i\dot{\hat{\kappa}}_i\ell_i+\sum_{i=1}^{N}\omega_i\mathcal{Q}_i(\hat{\kappa}_i-\bar{\kappa}_i)\dot{\hat{\kappa}}_i\\=&2\vartheta^T[(\hat{\kappa}+\ell)\Omega\mathcal{Q}\otimes(PA+A^TP)]\vartheta\\&-2\vartheta^T[(\hat{\kappa}+\ell)\mathcal{Q}\Omega\mathcal{L}_1\mathcal{Q}(\hat{\kappa}+\bar{\ell})\otimes PBB^TP]\tilde{\vartheta}\\&-2c\vartheta^T[(\hat{\kappa}+\ell)\mathcal{Q}\Omega\mathcal{L}_1\mathcal{Q}\otimes PB]\eta(\sigma)+2\vartheta^T[(\hat{\kappa}+\ell)\mathcal{Q}\Omega\mathcal{L}_1\otimes PB]\mathcal{P}\\&+\tilde{\vartheta}^T[(\hat{\kappa}+\ell-\bar{\kappa})\mathcal{Q}\Omega\otimes PBB^TP]\tilde{\vartheta}-\sum_{i=1}^{N}\omega_i\mathcal{Q}_i\nu_{\kappa i}(\hat{\kappa}_i+\ell_i-\bar{\kappa}_i)(\hat{\kappa}_i-1),\end{aligned} \tag{3.21}$$

where $\ell = diag\{\ell_1, \ell_2, \ldots, \ell_N\}$ and $\bar{\kappa} = diag\{\bar{\kappa}_1, \bar{\kappa}_2, \ldots, \bar{\kappa}_N\}$. Moreover, denote $\zeta = [\zeta_1^T(t), \zeta_2^T(t), \ldots, \zeta_N^T(t)]^T$, we have

$$\begin{aligned}
&-2\vartheta^T[(\hat{\kappa}+\ell)\mathcal{Q}\Omega\mathcal{L}_1\mathcal{Q}(\hat{\kappa}+\bar{\ell})\otimes PBB^TP]\tilde{\vartheta}\\
&=-2\vartheta^T[(\hat{\kappa}+\ell)\mathcal{Q}\Omega\mathcal{L}_1\mathcal{Q}(\hat{\kappa}+\ell)\otimes PBB^TP]\tilde{\vartheta}\\
&\quad+2\vartheta^T\big[(\hat{\kappa}+\ell)\mathcal{Q}\Omega\mathcal{L}_1\mathcal{Q}(\ell-\bar{\ell})\otimes PBB^TP\big]\tilde{\vartheta}\\
&=-2\vartheta^T[(\hat{\kappa}+\ell)\mathcal{Q}\Omega\mathcal{L}_1\mathcal{Q}(\hat{\kappa}+\ell)\otimes PBB^TP]\vartheta\\
&\quad-2\vartheta^T[(\hat{\kappa}+\ell)\mathcal{Q}\Omega\mathcal{L}_1\mathcal{Q}(\hat{\kappa}+\ell)\otimes PBB^TP]\zeta\\
&\quad+2\vartheta^T[(\hat{\kappa}+\ell)\mathcal{Q}\Omega\mathcal{L}_1\mathcal{Q}(\ell-\bar{\ell})\otimes PBB^TP]\tilde{\vartheta}. \qquad (3.22)
\end{aligned}$$

By Lemma 3.1, one obtains

$$\begin{aligned}
&-2\vartheta^T[(\hat{\kappa}+\ell)\mathcal{Q}\Omega\mathcal{L}_1\mathcal{Q}(\hat{\kappa}+\ell)\otimes PBB^TP]\vartheta\\
&=-\vartheta^T[(\hat{\kappa}+\ell)\mathcal{Q}(\Omega\mathcal{L}_1+\mathcal{L}_1^T\Omega)\mathcal{Q}(\hat{\kappa}+\ell)\otimes PBB^TP]\vartheta\\
&\le-\lambda_0\sum_{i=1}^N\mathcal{Q}_i^2(\hat{\kappa}_i+\ell_i)^2\vartheta_i^TPBB^TP\vartheta_i. \qquad (3.23)
\end{aligned}$$

By Young's inequality, for any constant $\varpi_1, \varpi_2 > 0$, and the eigenvalue of $\Omega\mathcal{L}_1\mathcal{L}_1^T\Omega$ is denoted by λ_Ξ, it follows

$$\begin{aligned}
&-2\vartheta^T[(\hat{\kappa}+\ell)\mathcal{Q}\Omega\mathcal{L}_1\mathcal{Q}(\hat{\kappa}+\ell)\otimes PBB^TP]\zeta\\
&\le\varpi_1\lambda_\Xi\sum_{i=1}^N\mathcal{Q}_i^2(\hat{\kappa}_i+\ell_i)^2\vartheta_i^TPBB^TP\vartheta_i\\
&\quad+\varpi_1^{-1}\sum_{i=1}^N\mathcal{Q}_i^2(\hat{\kappa}_i+\ell_i)^2\|B^TP\|^2\|\zeta_i\|^2. \qquad (3.24)
\end{aligned}$$

Moreover,

$$\begin{aligned}
&2\vartheta^T[(\hat{\kappa}+\ell)\mathcal{Q}\Omega\mathcal{L}_1\mathcal{Q}(\ell-\bar{\ell})\otimes PBB^TP]\tilde{\vartheta}\\
&\le\varpi_2\lambda_\Xi\sum_{i=1}^N\mathcal{Q}_i^2(\hat{\kappa}_i+\ell_i)^2\vartheta_i^TPBB^TP\vartheta_i\\
&\quad+\varpi_2^{-1}\tilde{\vartheta}^T[\mathcal{Q}^2(\ell-\bar{\ell})^2\otimes PBB^TP]\tilde{\vartheta}, \qquad (3.25)
\end{aligned}$$

Since $rank\big((\vartheta_i(t_k^i)+\vartheta_i(t))(\vartheta_i(t_k^i)+\vartheta_i(t))^T\big)=1$, $(\vartheta_i(t_k^i)+\vartheta_i(t))(\vartheta_i(t_k^i)+\vartheta_i(t))^T$ has only one non-zero eigenvalue. In light of Lemma 3.2, we can obtain $\lambda_{\max}[(\vartheta_i(t_k^i)+\vartheta_i(t))(\vartheta_i(t_k^i)+\vartheta_i(t))^T]=(\vartheta_i(t_k^i)+\vartheta_i(t))^T(\vartheta_i(t_k^i)+\vartheta_i(t))$. Then

$$
\begin{aligned}
&\tilde{\vartheta}^T[\mathcal{Q}^2(\ell-\bar{\ell})^2 \otimes PBB^TP]\tilde{\vartheta} \\
&= \sum_{i=1}^{N}(\vartheta_i^T(t)P\vartheta_i(t) - \vartheta_i^T(t_k^i)P\vartheta_i(t_k^i))^2\mathcal{Q}_i^2\vartheta_i^T(t_k^i)PBB^TP\vartheta_i(t_k^i) \\
&= \sum_{i=1}^{N}(\zeta_i^T(t)P\vartheta_i(t) + \vartheta_i^T(t_k^i)P\zeta_i(t))^2\mathcal{Q}_i^2\vartheta_i^T(t_k^i)PBB^TP\vartheta_i(t_k^i) \\
&= \sum_{i=1}^{N}[\zeta_i^T(t)P(\vartheta_i(t_k^i)+\vartheta_i(t))(\vartheta_i(t_k^i)+\vartheta_i(t))^TP\zeta_i(t)] \\
&\quad \times \mathcal{Q}_i^2\vartheta_i^T(t_k^i)PBB^TP\vartheta_i(t_k^i) \\
&\le \sum_{i=1}^{N}\lambda_{\max}[(\vartheta_i(t_k^i)+\vartheta_i(t))(\vartheta_i(t_k^i)+\vartheta_i(t))^T] \\
&\quad \times \mathcal{Q}_i^2\vartheta_i^T(t_k^i)PBB^TP\vartheta_i(t_k^i)\zeta_i^T(t)PP\zeta_i(t) \\
&= \sum_{i=1}^{N}\mathcal{J}_i\mathcal{Q}_i^2\zeta_i^T(t)PP\zeta_i(t).
\end{aligned}
\tag{3.26}
$$

In further,

$$
\begin{aligned}
&\tilde{\vartheta}^T[(\hat{\kappa}+\ell-\bar{\kappa})\mathcal{Q}\Omega \otimes PBB^TP]\tilde{\vartheta} \\
&\le (1+\varpi_3^{-1})\sum_{i=1}^{N}(\hat{\kappa}_i+\ell_i)\mathcal{Q}_i\omega_i\vartheta_i^T(t)PBB^TP\vartheta_i(t) \\
&\quad + (1+\varpi_3)\sum_{i=1}^{N}(\hat{\kappa}_i+\ell_i)\mathcal{Q}_i\omega_i\|B^T\|^2\|P\zeta_i(t)\|^2 \\
&\quad - \frac{1}{2}(1-\varpi_4)\sum_{i=1}^{N}\bar{\kappa}_i\mathcal{Q}_i\omega_i\vartheta_i^T(t)PBB^TP\vartheta_i(t) \\
&\quad + \frac{1}{2}\varpi_4^{-1}\sum_{i=1}^{N}\bar{\kappa}_i\mathcal{Q}_i\omega_i\|B^T\|^2\|P\zeta_i(t)\|^2 \\
&\quad - \frac{1}{2}\sum_{i=1}^{N}\bar{\kappa}_i\mathcal{Q}_i\omega_i\vartheta_i^T(t_k^i)PBB^TP\vartheta_i(t_k^i)
\end{aligned}
\tag{3.27}
$$

Note that if $\|B^T P \vartheta_i^T(t_k^i)\| \geq \chi_i, i = 1, 2, \ldots, N$.

$$\begin{aligned}
&\vartheta_i^T(t) P B \eta(B^T P \vartheta_i^T(t_k^i)) \\
&\quad = \frac{\vartheta_i^T(t) P B B^T P \vartheta_i(t_k^i)}{\|B^T P \vartheta_i(t_k^i)\|} \\
&\quad = \frac{(\vartheta_i(t_k^i) - \zeta_i)^T P B B^T P \vartheta_i(t_k^i)}{\|B^T P \vartheta_i(t_k^i)\|} \\
&\quad = \|B^T P \vartheta_i(t_k^i)\| - \frac{\zeta_i^T P B B^T P \vartheta_i(t_k^i)}{\|B^T P \vartheta_i(t_k^i)\|},
\end{aligned} \tag{3.28}$$

and

$$\begin{aligned}
&\vartheta_i^T(t) P B \eta(B^T P \vartheta_j(t_k^i)) \\
&\quad \leq \|B^T P \vartheta_i(t)\| \|\eta(B^T P \vartheta_j(t_k^i)\| \\
&\quad = \|B^T P \vartheta_i(t)\| \\
&\quad \leq \|B^T P \vartheta_i(t_k^i)\| + \|B^T P \zeta_i\|.
\end{aligned} \tag{3.29}$$

In light of (3.28) and (3.29), $\|\mathcal{P}_i\| \leq \mathcal{P}_i^S$, then we can obtain that

$$\begin{aligned}
&- 2\vartheta^T((\hat{\kappa} + \ell)\mathcal{Q}\Omega\mathcal{L}_1 \otimes PB)[c\mathcal{Q}\eta(\tilde{\delta}) - \mathcal{P}] \\
&\leq -2c \sum_{i=1}^{N} (\hat{\kappa}_i + \ell_i)\mathcal{Q}_i \omega_i \vartheta_i^T PB[a_{i0}\mathcal{Q}_i \eta(B^T P \vartheta_i(t_k^i)) \\
&\quad + \sum_{j=1}^{N} a_{ij}(\mathcal{Q}_i \eta(B^T P \vartheta_i(t_k^i)) - \mathcal{Q}_j \eta(B^T P \vartheta_j(t_k^i))] \\
&\quad + 2\sum_{i=1}^{N} (\hat{\kappa}_i + \ell_i)\mathcal{Q}_i \omega_i \vartheta_i^T PB[a_{i0}\mathcal{P}_i + \sum_{j=1}^{N} a_{ij}(\mathcal{P}_i - \mathcal{P}_j)] \\
&\leq 2\sum_{i=1}^{N} \mathcal{Q}_i \omega_i (\hat{\kappa}_i + \ell_i)(\mathcal{W}_{1i} c + \mathcal{W}_{3i})\|B^T P \vartheta_i(t_k^i)\| \\
&\quad + 2\sum_{i=1}^{N} \mathcal{Q}_i \omega_i (\hat{\kappa}_i + \ell_i)(\mathcal{W}_{2i} c + \mathcal{W}_{3i})\|B^T P \zeta_i\|,
\end{aligned} \tag{3.30}$$

where $\mathcal{W}_{1i} = \mathcal{L}_{ii}(\bar{\mathcal{Q}} - \mathcal{Q}_i) - a_{i0}\mathcal{Q}_i$, $\mathcal{W}_{2i} = \mathcal{L}_{ii}(\bar{\mathcal{Q}} + \mathcal{Q}_i) + a_{i0}\mathcal{Q}_i$, $\mathcal{W}_{3i} = (\mathcal{L}_{ii} + a_{i0})\mathcal{P}_i^S + \mathcal{L}_{ii}\mathcal{P}^S$, $\overline{\mathcal{Q}}$ and $\mathcal{P}^S$ are unknown positive constants, denoting the maximum of $\mathcal{Q}_i$ and $\mathcal{P}_i^S$, $i = 1, 2, \ldots, N$, respectively. Then,

$$
\begin{aligned}
&2\sum_{i=1}^{N}\mathcal{Q}_i\omega_i(\hat{\kappa}_i+\ell_i)(\mathcal{W}_{1i}c+\mathcal{W}_{3i})\|B^TP\vartheta_i(t_k^i)\| \\
&\le \frac{2}{\chi_i}\sum_{i=1}^{N}\mathcal{Q}_i\omega_i(\hat{\kappa}_i+\ell_i)(\mathcal{W}_{1i}c+\mathcal{W}_{3i})\|B^TP\vartheta_i(t_k^i)\|^2 \\
&\le \frac{2}{\chi_i}\sum_{i=1}^{N}(1+\varpi_5)\mathcal{Q}_i\omega_i(\hat{\kappa}_i+\ell_i)(\mathcal{W}_{1i}c+\mathcal{W}_{3i})\|B^TP\vartheta_i(t)\|^2 \\
&\quad+\frac{2}{\chi_i}\sum_{i=1}^{N}(1+\varpi_5^{-1})\mathcal{Q}_i\omega_i(\hat{\kappa}_i+\ell_i)(\mathcal{W}_{2i}c+\mathcal{W}_{3i})\|B^TP\zeta_i(t)\|^2. \qquad (3.31)
\end{aligned}
$$

Note that if $\|B^TP\vartheta_i^T(t_k^i)\| < \chi_i, i = 1, 2, \ldots, N$.

$$
\begin{aligned}
&-2\vartheta^T((\hat{\kappa}+\ell)\mathcal{Q}\Omega\mathcal{L}_1\otimes PB)[c\mathcal{Q}\eta(\tilde{\delta})-\mathcal{P}] \\
&\le 2\sum_{i=1}^{N}\mathcal{Q}_i\omega_i(\hat{\kappa}_i+\ell_i)(\mathcal{W}_{1i}c+\mathcal{W}_{3i})\chi_i \\
&\quad+2\sum_{i=1}^{N}\mathcal{Q}_i\omega_i(\hat{\kappa}_i+\ell_i)(\mathcal{W}_{2i}c+\mathcal{W}_{3i})\|B^TP\zeta_i\|. \qquad (3.32)
\end{aligned}
$$

Without loss of generality, assume that $\|B^TP\vartheta_i^T(t_k^i)\| \ge \chi_i, i = 1, 2, \ldots, l$, and $\|B^TP\vartheta_i^T(t_k^i)\| < \chi_i, i = l+1, l+2, \ldots, N$, where $2 \le l \le N-1$. By the following similar steps in the above two cases, it is not hard to derive that

$$
\begin{aligned}
&-2\vartheta^T((\hat{\kappa}+\ell)\mathcal{Q}\Omega\mathcal{L}_1\otimes PB)[c\mathcal{Q}\eta(\tilde{\delta})-\mathcal{P}] \\
&\le \frac{2}{\chi_i}\sum_{i=1}^{l}(1+\varpi_5)\mathcal{Q}_i\omega_i(\hat{\kappa}_i+\ell_i)(\mathcal{W}_{1i}c+\mathcal{W}_{3i})\|B^TP\vartheta_i(t)\|^2 \\
&\quad+\frac{2}{\chi_i}\sum_{i=1}^{l}(1+\varpi_5^{-1})\mathcal{Q}_i\omega_i(\hat{\kappa}_i+\ell_i)(\mathcal{W}_{2i}c+\mathcal{W}_{3i})\|B^TP\zeta_i(t)\|^2 \\
&\quad+2\sum_{i=l+1}^{N}\mathcal{Q}_i\omega_i(\hat{\kappa}_i+\ell_i)(\mathcal{W}_{1i}c+\mathcal{W}_{3i})\chi_i \\
&\quad+2\sum_{i=1}^{N}\mathcal{Q}_i\omega_i(\hat{\kappa}_i+\ell_i)(\mathcal{W}_{2i}c+\mathcal{W}_{3i})\|B^TP\zeta_i\|. \qquad (3.33)
\end{aligned}
$$

Note that

$$
\begin{aligned}
&2\sum_{i=1}^{N}(\mathcal{W}_{1i}c+\mathcal{W}_{3i})(\hat{\kappa}_i+\ell_i)\mathcal{Q}_i\omega_i\chi_i\\
&\leq\sum_{i=1}^{N}\frac{\nu_{\kappa_i}\mathcal{Q}_i\omega_i}{4}(\hat{\kappa}_i-1)^2+\sum_{i=1}^{N}\frac{1}{2\lambda_{\max}(P)}\mathcal{Q}_i\omega_i\ell_i^2+\sum_{i=1}^{N}\Xi_i\\
&\leq\sum_{i=1}^{N}\frac{\nu_{\kappa_i}\mathcal{Q}_i\omega_i}{4}(\hat{\kappa}_i-1)^2+\frac{1}{2}\vartheta^T(\mathcal{Q}\Omega\ell\otimes I_N)\vartheta+\sum_{i=1}^{N}\Xi_i,
\end{aligned}
\tag{3.34}
$$

where

$$
\Xi_i=2(\mathcal{W}_{1i}c+\mathcal{W}_{3i})\mathcal{Q}_i\omega_i\varsigma_i\chi_i+(\frac{4}{\nu_{\kappa_i}}+2\lambda_{\max}(P))(\mathcal{W}_{1i}c+\mathcal{W}_{3i})^2\chi_i^2\mathcal{Q}_i\omega_i.
\tag{3.35}
$$

It follows from (3.24)–(3.27) and (3.30)–(3.34) that

$$
\begin{aligned}
\dot{V}\leq&\vartheta^T(\hat{\kappa}+\ell)\Omega\mathcal{Q}\otimes\{(PA+A^TP)\\
&+[1+\varpi_3^{-1}+\frac{2}{\chi_i}(1+\varpi_5)(\mathcal{W}_{1i}c+\mathcal{W}_{3i})]PBB^TP\}\vartheta\\
&-\sum_{i=1}^{N}\{[\lambda_0-(\varpi_1+\varpi_2)\lambda_{\Xi}]\mathcal{Q}_i^2(\hat{\kappa}_i+\ell_i)^2\\
&+\frac{1}{2}(1-\varpi_4)\bar{\kappa}_i\mathcal{Q}_i\omega_i\}\vartheta_i^TPBB^TP\vartheta_i\\
&+\sum_{i=1}^{N}\bar{\kappa}_i\omega_i\mathcal{Q}_i\{\frac{\mathcal{Q}_i}{\beta_i\omega_i\bar{\kappa}_i\varpi_2}\beta_i\mathcal{J}_i+[\frac{1}{\beta_i\bar{\kappa}_i}\beta_i(1+\varpi_3)(\hat{\kappa}_i+\ell_i)\\
&+\frac{\mathcal{Q}_i}{\beta_i\omega_i\bar{\kappa}_i\varpi_1}\beta_i(\hat{\kappa}_i+\ell_i)^2+\frac{2}{\chi_i}\frac{(\mathcal{W}_{2i}c+\mathcal{W}_{3i})}{\beta_i\bar{\kappa}_i}\beta_i(1+\varpi_5^{-1})(\hat{\kappa}_i+\ell_i)\\
&+\frac{1}{2}\varpi_4^{-1}]\|B^T\|^2\}\|P\zeta_i(t)\|^2\\
&+2\sum_{i=1}^{N}\bar{\kappa}_i\omega_i\mathcal{Q}_i[\frac{(\mathcal{W}_{2i}c+\mathcal{W}_{3i})}{\beta_i\bar{\kappa}_i}\beta_i(\hat{\kappa}_i+\ell_i)]\|B^TP\zeta_i\|\\
&-\frac{1}{2}\sum_{i=1}^{N}\bar{\kappa}_i\omega_i\mathcal{Q}_i\|B^TP\vartheta_i(t_k^i)\|^2\\
&+\sum_{i=1}^{N}\frac{\nu_{\kappa_i}\mathcal{Q}_i\omega_i}{4}(\hat{\kappa}_i-1)^2+\frac{1}{2}\vartheta^T(\mathcal{Q}\Omega\ell\otimes I_N)\vartheta+\sum_{i=1}^{N}\Xi_i\\
&-\sum_{i=1}^{N}\omega_i\mathcal{Q}_i\nu_{\kappa_i}(\hat{\kappa}_i+\ell_i-\bar{\kappa}_i)(\hat{\kappa}_i-1).
\end{aligned}
\tag{3.36}
$$

Since, $\varpi_1 + \varpi_2 \in (0, (\lambda_0/\lambda_\Xi))$ and $\bar{\kappa}_i > 0$ to be sufficiently large such that $\bar{\kappa}_i > \max\left\{\mathcal{Y}_i, \frac{\mathcal{Q}_i}{\beta_i \omega_i \varpi_2}, \frac{1}{\beta_i}, \frac{\mathcal{Q}_i}{\beta_i \omega_i \varpi_1}, \frac{\mathcal{W}_{2i}}{\beta_i}, \frac{\mathcal{W}_{3i}}{\beta_i}\right\}$ holds, where $\mathcal{Y}_i = \frac{[2+\varpi_3^{-1}+\frac{2}{\chi_i}(1+\varpi_5)(\mathcal{W}_{1i}c+\mathcal{W}_{3i})]^2}{2(1-\varpi_4)\omega_i \mathcal{Q}_i[\lambda_0-(\varpi_1+\varpi_2)\lambda_\Xi]}$.

Note that

$$\begin{aligned} -(\hat{\kappa}_i - 1)(\hat{\kappa}_i - \bar{\kappa}_i) &= -(\hat{\kappa}_i - \bar{\kappa}_i)^2 - (\bar{\kappa}_i - 1)(\hat{\kappa}_i - \bar{\kappa}_i) \\ &\leq -\frac{1}{2}(\hat{\kappa}_i - \bar{\kappa}_i)^2 + \frac{1}{2}(\bar{\kappa}_i - 1)^2, \end{aligned} \tag{3.37}$$

and

$$\begin{aligned} -(\hat{\kappa}_i - 1)(\hat{\kappa}_i - \bar{\kappa}_i) &= -(\hat{\kappa}_i - 1)^2 - (\bar{\kappa}_i - 1)(\hat{\kappa}_i - \bar{\kappa}_i) \\ &\leq -\frac{1}{2}(\hat{\kappa}_i - 1)^2 + \frac{1}{2}(\bar{\kappa}_i - 1)^2. \end{aligned} \tag{3.38}$$

Furthermore

$$\begin{aligned} &-\sum_{i=1}^{N} \omega_i \mathcal{Q}_i \nu_{\kappa_i} (\hat{\kappa}_i - \bar{\kappa}_i)(\hat{\kappa}_i - 1) \\ &= \sum_{i=1}^{N} \frac{\omega_i \mathcal{Q}_i \nu_{\kappa_i}}{2} (\bar{\kappa}_i - 1)^2 - \frac{\omega_i \mathcal{Q}_i \nu_{\kappa_i}}{4} [(\hat{\kappa}_i - \bar{\kappa}_i)^2 + (\hat{\kappa}_i - 1)^2]. \end{aligned} \tag{3.39}$$

Then, trigger function (3.10) ensures that

$$\begin{aligned} \dot{V} \leq &- \delta V - \vartheta^T ((\hat{\kappa} + \ell)\Omega \mathcal{Q} \otimes \left(\frac{1}{2} I_N - \delta P\right) \vartheta - \sum_{i=1}^{N} \left(\frac{\nu_{\kappa_i}}{4} - \frac{\delta}{2}\right) \omega_i \mathcal{Q}_i \left(\hat{\kappa}_i - \bar{\kappa}_i\right)^2 \\ &+ \frac{1}{2}(\bar{\kappa}_i - 1)^2 \sum_{i=1}^{N} \omega_i \nu_{\kappa_i} \mathcal{Q}_i + \sum_{i=1}^{N} \Xi_i + \sum_{i=1}^{N} \bar{\kappa}_i \omega_i \rho_i e^{-o_i t}. \end{aligned} \tag{3.40}$$

Moreover, based on the condition $0 < \delta \leq \min\{\frac{\nu_{\kappa_i}}{2}, \frac{1}{2\lambda_{\max}(P)}\}$ and (3.12), such that (3.40) satisfies the following inequality:

$$\dot{V} \leq -\delta V + \frac{1}{2}(\bar{\kappa}_i - 1)^2 \sum_{i=1}^{N} \omega_i \nu_{\kappa_i} \mathcal{Q}_i + \sum_{i=1}^{N} \Xi_i + \sum_{i=1}^{N} \bar{\kappa}_i \omega_i \rho_i e^{-o_i t}. \tag{3.41}$$

V exponentially converges to a bounded set

$$\mathcal{D} \triangleq \left\{ \vartheta, \hat{\kappa} : V \leq \frac{1}{\delta} \sum_{i=1}^{N} [\frac{1}{2}(\bar{\kappa}_i - 1)^2 \omega_i \nu_{\kappa_i} \mathcal{Q}_i + \Xi_i] \right\}. \tag{3.42}$$

Therefore ϑ, $\hat{\kappa}$ are uniformly ultimately bounded and ϑ exponentially converges to the residual set $\mathcal{D}$. ■

Remark 3.1 Note that the definition of the triggering state in [4, 5] is $\widetilde{x}_i(t) = e^{A(t-t_k^i)} x_i(t_k^i), \forall i \in [t_k^i, t_{k+1}^i)$. The controller u_i in the above work involves $\widetilde{x}_i - \widetilde{x}_j$. In general, $t_k^i \neq t_k^j, i \neq j$. Therefore, the triggering state defined in the controller in exponential form needs to know the absolute state of the corresponding agent. In contrast, the controller in (3.15) and the trigger condition (3.10) only need the relative state between agents.

Remark 3.2 By introducing dynamic coupling gains $\hat{\kappa}_i$ and $\bar{\ell}_i$, the coupling gains are updating using relative errors $\vartheta_i(t_k^i)$. Traditionally, the dynamic coupling gains are constant and do not use an adaptive mechanism. Determining these scalars requires the Laplacian or the scale of the communication graph $\mathcal{G}$. It is worth mentioning that the parameter $\bar{\kappa}_i, i = 1, 2, \ldots, N$ associated with the global information in this chapter are sufficiently large positive constants included in (3.20), not in the control algorithm.

Remark 3.3 In existing work [6], fully distributed event-triggered algorithms under directed graphs require multiple event monitoring mechanisms, which is certainly a burden on the hardware. As Lemma 3.2 is applied to (3.26), a novel event-triggered mechanism is proposed in this chapter. The consensus problem under general directed graphs can be solved using only one event-triggered condition.

Remark 3.4 The proposed ETFT protocol uses boundary layer techniques to successively approximate nonsmooth functions to overcome the undesirable chatting effect of the nonsmooth of the static protocol in [7]. By (3.42), to ensure that the consensus error ϑ converges progressively to a satisfactory vicinity of zero, we can choose a relatively small χ_i, ν_{κ_i}. It is important to mention that the design of the ETFT-based continuous controller (3.15) was partly inspired by Cheng and Li [5]. However, the design process is quite different due to the heterogeneity of multiplicative faults and the asymmetry of the topology graph.

Theorem 3.2 *Considering the system (3.3), if Assumption 3.1 holds, the triggered rule is designed as (3.10), the controller is given as (3.15), and then the Zeno phenomenon can be excluded.*

Proof Taking the time derivative on $\|\zeta_i(t)\|$, one has:

$$D^+ \|\zeta_i(t)\| = \frac{d}{dt}\sqrt{\zeta_i^T(t)\zeta_i(t)} \leq \|\dot{\zeta}_i(t)\|. \tag{3.43}$$

From the definition of $\epsilon_i(t)$, for $t \in [t_k^i, t_{k+1}^i)$, we have

$$\begin{aligned}\dot{\zeta}_i(t) = -\dot{\vartheta}(t) =& A\zeta_i(t) - A\vartheta_i(t_k^i) + (\hat{\kappa}_i + \bar{\ell}_i)\mathcal{Q}_i BB^T P\vartheta_i(t_k^i) \\ &+ c\mathcal{Q}_i B\eta(\tilde{\sigma}_i) - B\mathcal{P}_i.\end{aligned} \tag{3.44}$$

It follows from Theorem 3.1 that $\vartheta_i(t)$ is bounded. It's easy to see that $\vartheta_i(t_k^i)$ is also bounded.

Denote

$$\begin{aligned}\sigma_i = &\max_{t\in[t_k^i, t_{k+1}^i]} \|A\vartheta_i(t_k^i)\| + \|(\hat{\kappa}_i + \bar{\ell}_i)\mathcal{Q}_i BB^T P\vartheta_i(t_k^i)\| \\ &+ c\|\mathcal{Q}_i B\eta(\tilde{\sigma}_i)\| + \|B\mathcal{P}_i\|.\end{aligned} \tag{3.45}$$

We can follow from (3.43)–(3.45) that

$$\frac{d\|\zeta_i(t)\|}{dt} \le \|A\|\|\zeta_i(t)\| + \sigma_i. \tag{3.46}$$

Note that, $\zeta_i(t_k^i) = 0$. By Comparison Principle, for $t \in [t_k^i, t_{k+1}^i]$, $\|\zeta_i(t)\|$ satisfies the following inequality:

$$\|\zeta_i(t)\| = \frac{\sigma_i}{\|A\|}(e^{\|A\|(t-t_k^i)} - 1). \tag{3.47}$$

The trigger function (3.10) satisfies $f_i(t) \le 0$, if the following conditions is established:

$$\|\zeta_i\| \le \frac{-2\zeta(\hat{\kappa}_i + \ell_i)\|B^T\| + \sqrt{\Delta}}{2\mathcal{M}_{1i}\|P\|} \triangleq \zeta_i^*, \tag{3.48}$$

where

$$\Delta =(2\zeta(\hat{\kappa}_i + \ell_i)\|B\|)^2 + 4\mathcal{M}_{1i}(\frac{1}{2}\vartheta_i^T(t_k^i)PBB^T P\vartheta_i(t_k^i) + \rho_i e^{-o_i t}). \tag{3.49}$$

Then, solve the following inequality:

$$\frac{\sigma_i}{\|A\|}(e^{\|A\|\tau_k^i} - 1) \ge \zeta_i^*, \tag{3.50}$$

a lower bound τ_k^i of $t_{k+1}^i - t_k^i$ can be obtained.

Based on (3.46), (3.47) and (3.50), it yields that

$$t_{k+1}^i - t_k^i \ge \tau_k^i \ge \frac{1}{\|A\|}ln(1 + \frac{\|A\|}{\sigma_i}\zeta_i^*). \tag{3.51}$$

Note that τ_k^i always exists and is strictly positive. Then, the Zeno phenomenon has been excluded. This completes the proof. ■

As a result, the proposed control strategy ensures that the voltages of all DGs are restored to the reference value.

Different from the classical MASs algorithms [4] based on an undirected graph, in this chapter, the measurement error $\zeta_i(t)$ is defined as (3.11) to deal with multiplicative faults. As can be seen from $\vartheta_i(t)$, the trigger mechanism (3.10) requires DGs to continuously monitor the information of its neighbors, which can lead to a waste of network communication resources. When the fault factors are known or fixed constants, using discrete state quantities to predict ϑ can eliminate the limitation of continuous communication. For this purpose, we have

$$\begin{aligned}\dot{\vartheta}_i &= \sum_{j=1}^{N} \mathcal{L}_{ij}\left(\dot{x}_j - \dot{x}_0\right) \\ &= A\vartheta_i + B\mathcal{L}_{ii}\left(\mathcal{Q}_i u_i + \mathcal{P}_i\right) - B\sum_{j=1}^{N} a_{ij}\left(\mathcal{Q}_j u_j + \mathcal{P}_j\right).\end{aligned} \tag{3.52}$$

For $t \in [t_k^i, t_{k+1}^i)$, we can obtain (3.53). Therefore, the triggering function (3.10) can avoid the continuous communication of adjacent DGs. The predicted consensus error $\vartheta_i(t)$ in (3.53) is only related to the states of the neighboring agents, so the algorithm proposed in this chapter is still fully distributed.

$$\begin{aligned}\vartheta_i(t) =& \mathcal{L}_{ii} e^{At} \int_{t_k^i}^{t} e^{-A\tau} B\mathcal{Q}_i \Big[-(\hat{\kappa}_i + \bar{\ell}_i) B^T P \vartheta_i(t_k^i) - c\eta(B^T P \vartheta_i(t_k^i)) \Big] d\tau \\ & - e^{At} \int_{t_k^i}^{t} e^{-A\tau} B \sum_{j=1}^{N} a_{ij} \Big[-\mathcal{Q}_j(\hat{\kappa}_j + \bar{\ell}_j) B^T P \vartheta_j(t_k^J) - c\mathcal{Q}_j \eta(B^T P \vartheta_j(t_k^i)) \Big] d\tau \\ & + e^{A(t-t_k^i)} \vartheta_i(t_k^i).\end{aligned} \tag{3.53}$$

3.4 Simulation Results

In this section, to demonstrate the effectiveness of the proposed adaptive ETFT control algorithm, various scenarios are simulated for an isolated MG in a MATLAB/Simulink software environment. The minimal communication swarm is shown in Fig. 3.1. The islanded MG test is a combination of four inverter-based DGs. The simulation results are extracted by modeling the dynamical equations of MG in Matlab. Table 3.1 provides the DGs, lines, and loads parameters.

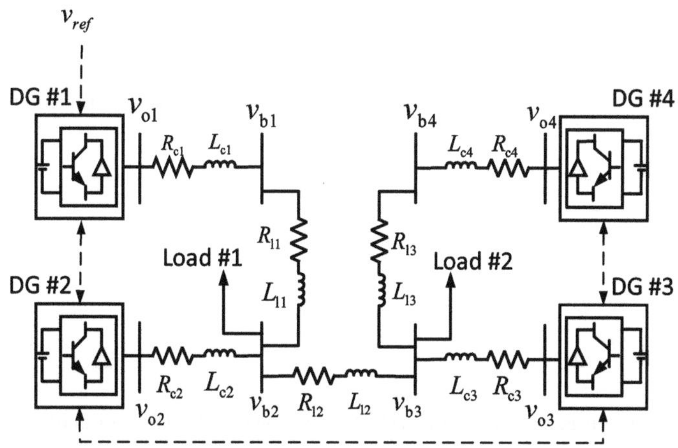

Fig. 3.1 The example MG test system and considered communication structure

Table 3.1 Specifications of the microgrid test system

DGs	DG#1 & DG#2 & DG#3 & DG#4	
	m_P	1.5×10^{-5}
	n_Q	2×10^{-4}
	R_c	0.35 Ω
	L_c	1.847 mH
	K_{PV}	0.2013
	K_{IV}	480
	K_{PC}	6.72
	K_{IC}	3360
Lines	Line1&Line3	Line2
	$R_{l1} = R_{l2} = 0.2\Omega$	$R_{l1} = 0.35\Omega$
	$L_{l1} = L_{l2} = 0.318mH$	$L_{l1} = 1.847mH$
RL loads	Load #1 & Load #2	
	P (per phase)	Q (per phase)
	50 kW	12 kVAr

3.4.1 Case 1: Effectiveness Evaluation

This section discusses the effectiveness of the proposed ETFT control algorithm.

Solving the ARE (3.12) gives a solution

$$P = \begin{bmatrix} 1.7321 & 1.0000 \\ 1.0000 & 1.7321 \end{bmatrix},$$

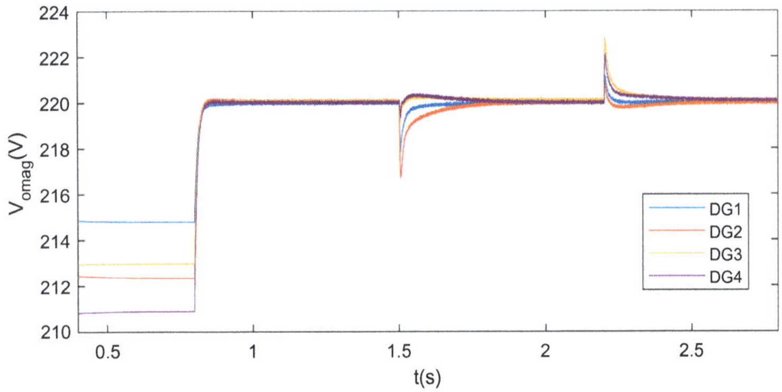

Fig. 3.2 Output voltage magnitudes for each DG applying the EDFTC algorithms in Case 1

The control parameters of the adaptive ETFT control algorithm are chosen as $c = 1$, $\chi_i = 0.1$, $\varpi_3 = \varpi_4 = \varpi_5 = 1$, $\beta_i = 1e-5$, $\gamma_i = 0.05$, $\nu_{\kappa_i} = 1, i = 1, 2, \ldots, N$, $\rho_1 = 15$, $\rho_2 = \rho_3 = \rho_4 = 12$.

The PLOE faults and bias faults in the DGs are chosen as follows: $\mathcal{Q}_1 = 0.4\cos(5t-3)+0.7$, $\mathcal{Q}_2 = 0.1\cos(3t-3)+0.1$,

$$\mathcal{Q}_3 = \begin{cases} 0.8\sin(1t-3)+0.1, \mathit{if}\ t > 1,4\,\text{s} \\ 0.4, \qquad\qquad \mathit{if}\ 0.8 < t \le 1.4\,\text{s}, \end{cases} \quad \mathcal{Q}_4 = 0.09, \mathcal{P}_1 = 4\cos(6t-7)+4,$$

$\mathcal{P}_2 = 3\sin(t-3)+1$, $\mathcal{P}_3 = 0.1\cos(2t-10)+0.4$ and $\mathcal{P}_4 = 2\cos(2t-2)+0.21$.

To show the tracking performance and fault tolerance of the proposed adaptive ETFT control algorithm, the secondary control is intentionally deactivated at the beginning and only the primary control is applied. At $t = 0.8$ s, the adaptive ETFT control algorithm is applied in secondary layer. At $t = 1.5$ s, the load is decreased by 50%. At $t = 2.2$ s, the load is increased load to its nominal value.

As it can be seen in Fig. 3.2, the output voltages of the four DGs deviate somewhat from the nominal values at the beginning of the simulation. Afterwards, the proposed ETFT control algorithms are applied at $t = 0.8s$, all terminal voltage amplitudes return to the reference value within $0.1s$. Next, the step load change scenario is investigated by decreasing load at $t = 1.5s$ and increasing load at $t = 2.2s$ (Fig. 3.3).

The triggering instants for each DG applying the algorithms are shown in Fig. 3.5, which shows no Zeno phenomenon occurs.

Therefore, the proposed ETFT control algorithm and the adaptive ETFT control algorithm can not only restore the voltage to the reference value in case of actuator faults, but also reduce the energy in communication.

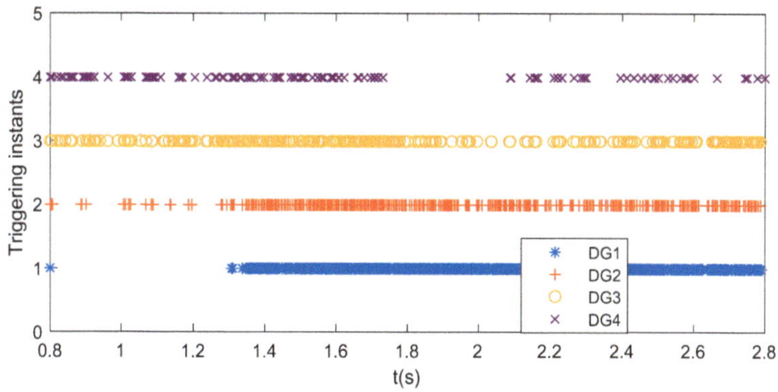

Fig. 3.3 Output voltage magnitudes for each DG applying the EDFTC algorithms in Case 1

3.4.2 *Case 2: Changes in Fault and PIPO Capability*

This section discusses the sensitivity of the proposed adaptive algorithms to changes in fault parameters and the PIPO capability.

The control parameters of the adaptive ETFT control algorithm are chosen as $\rho_1 = 18$, $\rho_2 = \rho_3 = \rho_4 = 14$, other control parameters are the same as case 1. The PLOE faults and bias faults in the DGs are chosen as follows: $\mathcal{Q}_1 = 0.28\cos(5t - 1) + 0.7$, $\mathcal{Q}_2 = 0.1\cos(3t - 3) + 0.1$, $\mathcal{Q}_3 = 0.7sin(1t - 3) + 0.2$, $\mathcal{Q}_4 = 0.09$, $\mathcal{P}_1 = 2\cos(3t - 3) + 4$, $\mathcal{P}_2 = 1.4\sin(5t - 7) + 4$, $\mathcal{P}_3 = 5\cos(2t - 10) + 0.4$, $\mathcal{P}_4 = 3\cos(3t - 2) + 3$. Similarly, at $t = 0.8$ s, the secondary control is activated. Differently, at $t = 1.5$ s, DG#4 is withdrawn from operation and reconnected at $t = 2.2$ s.

Figure 3.4 reveals that the proposed ETFT control algorithms have good tracking performance and fault tolerance even after changes in faults. Moreover, the proposed control strategies have good responsiveness and robustness to the disconnection and reconnection of the DG units. Although there are some transients after DG#4 disconnection and reconnection, it achieves the PIPO function of MG. The triggering instants for each DG applying the algorithms are shown in Fig. 3.5.

3.5 Conclusions

The feedback linearization method is applied to transform the secondary control problem of an isolated MG system into a tracking consensus problem of MAS. Then, the secondary voltage control with adaptive ETFT control is proposed to restore the output voltage amplitude of all DGs to the reference value. The algorithm relies only on the information related to the adjacent DGs. It eliminates the need for global infor-

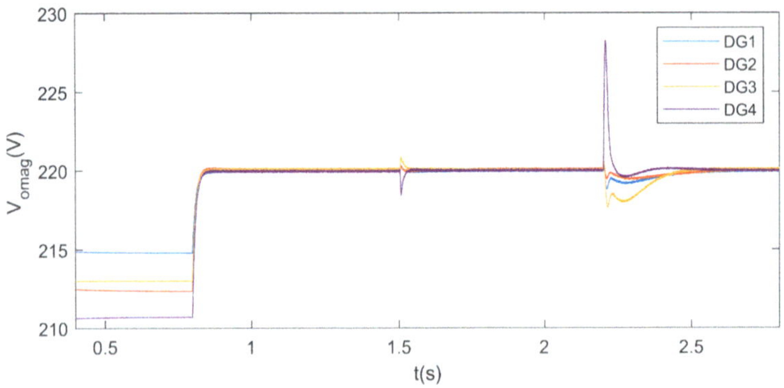

Fig. 3.4 Output voltage magnitudes for each DG applying the EDFTC algorithms in Case 2

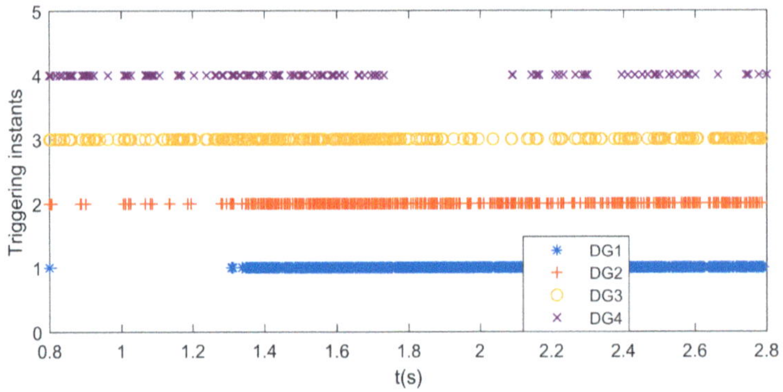

Fig. 3.5 The triggering instants for each DG applying the adaptive EDFTC algorithms in Case 2

mation and fault boundaries through adaptive control. Moreover, a consensus error prediction model is proposed to eliminate the continuous communication between DGs. A Matlab/Simulink simulation system verified the effectiveness of the proposed secondary control strategy.

References

1. Bidram, A., Davoudi, A., Lewis, F.L., Guerrero, J.M.: Distributed cooperative secondary control of microgrids using feedback linearization. IEEE Trans. Power Syst. **28**(3), 3462–3470 (2013)
2. Qu, Z.: Cooperative control of dynamical systems: applications to autonomous vehicles. Springer, London (2009)
3. Zhang F.: Matrix theory: basic results and techniques (2011)
4. Cheng, B., Li, Z.: Designing fully distributed adaptive event-triggered controllers for networked linear systems with matched uncertainties. IEEE Trans. Neural Netw. Learn. Syst. **30**(12), 3645–3655 (2019)

5. Cheng, B., Li, Z.: Fully distributed event-triggered protocols for linear multi-agent networks. IEEE Trans. Autom. Control **64**(4), 1655–1662 (2019)
6. Li, X., Sun, Z., Tang, Y., Karimi, H.R.: Adaptive event-triggered consensus of multiagent systems on directed graphs. IEEE Trans. Autom. Control **66**(4), 1670–1685 (2021)
7. Luo, S., Ye, D.: Adaptive double event-triggered control for linear multi-agent systems with actuator faults. IEEE Trans. Circ. Syst. I: Regular Papers **66**(12), 4829–4839 (2019)

Chapter 4
Output-Feedback Consensus Fault-Tolerant Control With Intermittent Communication to Achieve Load Voltage Regulation

Abstract This chapter proposes an observer-based event-driven fault-tolerant (OBEDFT) secondary control strategy for AC microgrids (MGs) to achieve load voltage regulation. First, the input-output feedback linearization method transforms the voltage regulation issue into an output feedback tracking problem for linear multiagent systems (MASs) with nonlinear dynamics. This transformation provides the necessary preprocessing for load voltage regulation. Then, an OBEDFT secondary control protocol that considers full-state immeasurability is proposed. The actuators of distributed generators (DGs) may experience partial loss of effectiveness (PLOE) and bias faults, and these fault parameters may be heterogeneous and time-varying. The protocol introduces adaptive techniques to avoid information related to fault parameters while using event-driven mechanisms to achieve discrete measurements of neighboring DG. Additionally, the protocol uses boundary layers to construct smooth controllers that prevent the chattering effect caused by nonsmooth controllers. Finally, simulation results confirm the effectiveness of this load voltage regulation strategy.

Keywords AC microgrid (MG) · Event-triggered control · Fault-tolerant control · Multiagent system (MAS) · Voltage regulation

4.1 Introduction

In this chapter, we proposes a load voltage regulation strategy for AC MGs using distributed OBEDFT tracking control. The load voltage regulation issue is transformed into a distributed output feedback tracking problem for linear MASs with nonlinear dynamics and full-state immeasurability. This transformation provides the necessary pre-processing for designing the OBEDFT tracking control protocol for load voltage regulation to the reference value. This research solves the complex decoupling problem of the heterogeneous multiplicative fault parameters and the combined measurement error $e_i(t)$. We address the tracking problem by developing a discrete OBEDFT protocol that avoids the combined measurement error $e_i(t)$ and thus allows communication between DGs at discrete trigger moments. A novel design process

J. Sun et al., *Building a Low-Carbon Future*, Studies in Systems, Decision and Control 506, https://doi.org/10.1007/978-3-031-43498-3_4

for an adaptive OBEDFT control protocol with a smooth controller is proposed. The control protocol does not require quantitative information about the magnitude, frequency, and type (whether constant, timevarying, periodic, etc.) of PLOE faults and bias faults present in the actuator. It only needs to guarantee the boundedness of the fault signal. By introducing the concepts of adaptive and boundary layers in the protocol, we enable the control protocol to overcome the limitations of fault factors and the undesirable chattering effects encountered by non-smooth controllers.

4.1.1 MG System Large-Signal Dynamical Model

The considered MG contains N DGs, and $\mathcal{G}$, which is designed later, represents the topology diagram of the upper layer communication between DGs. In Fig. 4.1, the inverter-based MGs' dynamic model includes droop control inner-loop voltage and current power, LC filter, and line model. Also, the designed OBEDFT secondary control strategy is shown in Fig. 4.1.

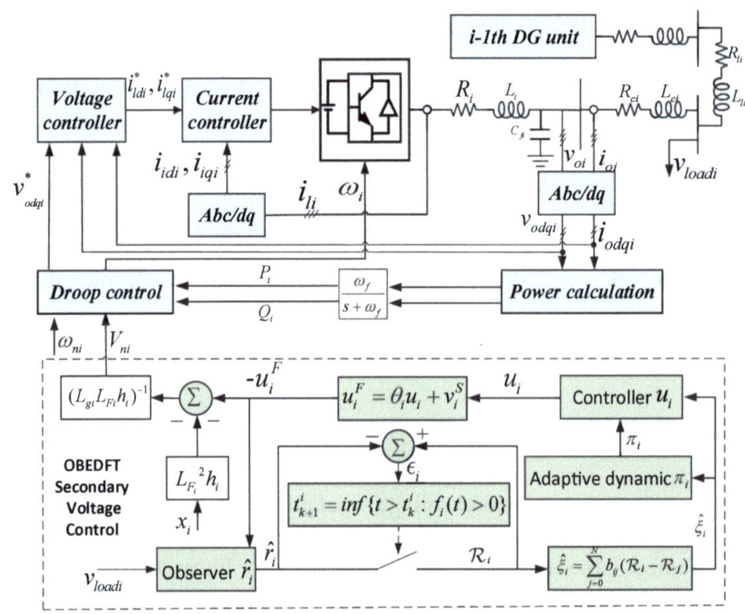

Fig. 4.1 Block diagram of an inverter-based DG and OBEDFT secondary voltage regulation control

The nonlinear dynamic model of the AC MG is the collection in a compact form as

$$\dot{x}_i = f_i(x_i) + k_i(x_i)D_i + g_i(x_i)u_i$$
$$\mathcal{Y}_i = h_i(x_i), \tag{4.1}$$

where the state vector is

$$x_i = [\delta_i, P_i, Q_i, \phi_{v_{di}}, \phi_{v_{qi}}, \phi_{i_{di}}, \phi_{i_{qi}}, i_{ldi}, i_{lqi}, v_{odi}, v_{oqi}, i_{odi}, i_{oqi}]^T, \tag{4.2}$$

and $D_i = [\omega_{com}, v_{bdi}, v_{bqi}]$. Detailed expressions for $f_i(x_i)$, $g_i(x_i)$, $k_i(x_i)$, and $h_i(x_i)$ can be extracted from [1]. The output is set to $\mathcal{Y}_i = v_{odi}$.

Then, by feedback linearization, we have

$$\ddot{\mathcal{Y}}_i = L_{F_i}^2 h_i + L_{g_i} L_{F_i} h_i V_{ni} = u_i, \tag{4.3}$$

where $F_i = f_i(x_i) + k_i(x_i)D_i$, $L_{F_i} h_i = \frac{\partial h_i}{\partial x_i} F_i$ and $L_{F_i}^2 h_i = L_{F_i}(L_{F_i} h_i) = \frac{\partial L_{F_i} h_i}{\partial x_i} F_i$ are the Lie derivatives of h_i along F_i.

Because of $v_{load_i} = \mathcal{Y}_i + v_{li}$, where v_{load_i}, v_{li} denote the load voltage and the line voltage between the DG output and the load side, respectively. Then, we can obtain a second order linear system as follows:

$$\ddot{v}_{load_i} = \ddot{\mathcal{Y}}_i + \ddot{v}_{li} = u_i + \ddot{v}_{li}. \tag{4.4}$$

Then, for any DG, r_i is translated from (4.4) to

$$\begin{cases} \dot{r}_i(t) = Ar_i(t) + B(u_i + \ddot{v}_{li}), \\ y_i(t) = Cr_i(t), i = 1, \ldots, N, \end{cases} \tag{4.5}$$

where $r_i = [v_{load_i}, \dot{v}_{load_i}]^T$, $A = \begin{bmatrix} 0 & 1 \\ 0 & 0 \end{bmatrix}$, $B = [0, 1]^T$, $C = [1, 0]$.

Remark 4.1 In this chapter, we address the load voltage v_{load_i} regulation problem for AC MGs. After feedback linearization process, the load voltage regulation problem is transformed into a distributed output feedback tracking problem with linear MASs, where the system contains nonlinear dynamics.

The reference voltage dynamics can be expressed as

$$\dot{r}_0(t) = Ar_0(t), \tag{4.6}$$

where $r_0 = [v_{ref}, 0]^T$.

4.1.2 Notation and Graph Theory

Upper layer communication topology diagram $\mathcal{G}$ consists of the node set $\gimel = \{\gimel_0, \gimel_1, \ldots, \gimel_N\}$, the edge set $\mathcal{E} \subseteq \gimel \times \gimel$ and a weighted adjacency matrix $\mathcal{A} = [a_{ij}] \in \mathcal{R}^{(N+1)\times(N+1)}$, where $a_{ij} \geq 0$. $a_{ij} > 0$ when and only when $(\gimel_i, \gimel_j) \in \mathcal{E}$. This also implies that $\gimel_i$ can receive information from $\gimel_j$, and vice versa, then $a_{ij} = 0$. Moreover, the Laplacian matrix $\mathcal{L}$ is defined as $\mathcal{L}_{ii} = \sum_{j=0}^{N} a_{ij}$ and $\mathcal{L}_{ij} = -a_{ij}$, where $i \neq j$.

4.1.3 Fuzzy Logic Systems

Since the $\ddot{v}_{li}, i = 1, 2, \ldots, N$ in (4.4) are unknown nonlinear dynamics, fuzzy logic systems (FLSs) are used to approximate these unknown nonlinear dynamics. The properties of FLSs are represented by the following lemma.

Lemma 4.1 *[2] Let the domain of continuous function $g(x)$ be a compact set Ω. Then, for any $\varepsilon > 0$, there exists a FLS (7) such that*

$$\sup_{x\in\Omega} |\varphi^T(x)W - g(x)| \leq \varepsilon. \tag{4.7}$$

From Lemma 4.1, we approximate the unknown nonlinear function $\ddot{v}_{li}$ as

$$\ddot{v}_{li} = \varphi_i^T(r_i)W_i + o_i(t), \tag{4.8}$$

where $W_i = [W_{i1}, W_{i2}, \ldots, W_{im}]^T$ is the unknown parameter. $\varphi_i^T(r_i) = [\varphi_{i1}, \varphi_{i2}, \ldots, \varphi_{im}]$ is the fuzzy basis function. In addition, o_i with $\|o_i(t)\| \leq \bar{o}_i$ is a time-varying function. $\bar{o}_i$ is an unknown constant.

Lemma 4.2 *[3] The basis function in (4.8) satisfies $\varphi_i(\cdot)\varphi_i^T(\cdot) \leq lI$.*

4.2 Distributed OBEDFT Load Voltage Regulation Control

4.2.1 Nodes Dynamics and Fault Model

To investigate the fault-tolerant control problem, the fault model must be established. Here, we consider actuator faults when PLOE coexists with bias faults. The actual output of the actuator is represented as follows:

$$u_i^F = \theta_i(t)u_i + v_i^S, \tag{4.9}$$

where $\theta_i(t)$, v_i^S are fault parameter representing PLOE fault and biased fault severity of the ith DG, respectively.

Assumption 4.1 $\theta_i(t)$ with $0 < \underline{\theta}_i \leq \theta_i(t) \leq \overline{\theta}_i, i = 1, 2, \ldots, N$ are time-varying. $\underline{\theta}_i$ and $\overline{\theta}_i$ are unknown positive constants that stand for the lower and upper bounds of $\theta_i(t)$.

Assumption 4.2 $\|v_i^S\| \leq \varsigma_i$ with $\varsigma_i > 0$ being an unknown constant.

Remark 4.2 Assumptions 4.1 and 4.2 guarantee the boundedness of the considered actuator PLOE and deviation faults. However, all information related to the fault parameters is unknown, including quantitative information on boundaries, frequencies, and types (whether they are constant, time-varying, periodic, etc.). It is well known that the fault parameters can be different for different DGs. These assumptions are much more relaxed and more consistent with the fault parameter conditions than in [4] for practical applications.

Due to (4.8), the following state equation can represent the dynamics of each DG:

$$\begin{cases} \dot{r}_i(t) = Ar_i(t) + B(u_i^F + \varphi_i^T(r_i)W_i + o_i), \\ y_i(t) = Cr_i(t). \end{cases} \tag{4.10}$$

Since it is difficult to obtain the value of $\dot{v}_{loadi}$ in practice, the designed control protocol uses only y_i and not r_i. To estimate the system state, the following state observer for ith DG is designed:

$$\begin{cases} \dot{\hat{r}}_i(t) = A\hat{r}_i(t) + B(u_i^F + \varphi_i^T(\hat{r}_i)\hat{W}_i) - L(\hat{y}_i - y_i), \\ \hat{y}_i(t) = C\hat{r}_i(t), i = 1, 2, \ldots, N, \end{cases} \tag{4.11}$$

where $\hat{r}_i(t)$ is the estimation of $r_i(t)$, $\hat{y}_i(t)$ is the observer output vector, L is the constant matric to be specified, and $\hat{W}_i$ is designed later.

4.2.2 *Design of Distributed OBEDFT Secondary Voltage Control Protocol*

The distributed OBEDFT load voltage regulation strategy aims to regulate the load voltage v_{load_i} to the reference value v_{ref}. In this section, the distributed OBEDFT secondary voltage control protocol is designed.

Assumption 4.3 The communication graph $\mathcal{G}$ contains a directed spanning tree, where the leader is the root node and the communication between followers is bidirectional, but the weights may be asymmetric.

Then, we can obtain

$$\mathcal{L} = \begin{bmatrix} 0 & 0_{1\times N} \\ \mathcal{L}_{21} & \mathcal{L}_1 \end{bmatrix}, \tag{4.12}$$

where $\mathcal{L}_{21} \in \mathcal{R}^{N\times 1}$, $\mathcal{L}_1 \in \mathcal{R}^{N\times N}$. The real part of each eigenvalue of $\mathcal{L}_1$ is positive. It is easy to verify that $\mathcal{L}_1$ is a nonsingular M-matrix.

Suppose the interaction graph $\bar{\mathcal{G}}$ possesses the same subgraphs as $\mathcal{G}$, with the difference that the communication between leaders and followers in $\bar{\mathcal{G}}$ is bidirectional. Then, the Laplacian matrix of $\bar{\mathcal{G}}$ can be expressed as

$$\hat{\mathcal{L}} = \begin{bmatrix} \mathcal{L}_0 & \mathcal{L}_{12} \\ \mathcal{L}_{21} & \mathcal{L}_1 \end{bmatrix}, \tag{4.13}$$

where $\mathcal{L}_0 = \sum_{j=1}^{N} \bar{a}_j$, $\mathcal{L}_{12} = [-\bar{a}_1, -\bar{a}_2, \ldots, -\bar{a}_N] \in \mathcal{R}^{1\times N}$, $\mathcal{R}^{n\times m}$ denotes the set of all $n \times m$ real matrices, $\bar{a}_j$, $j = 1, 2, \ldots, N$ are designed constants satisfying $\bar{a}_j > 0$ if the leader directly communicates to follower ith DG in graph $\mathcal{G}$, and $\bar{a}_j = 0$, otherwise. It is easy to see that the $\bar{\mathcal{G}}$ becomes strongly connected.

Then, using Lemma 4.2 we get

$$\hat{\Psi} = \begin{bmatrix} 2\mathcal{L}_0 b_0 & b_0\mathcal{L}_{12} + \mathcal{L}_{21}^T \hat{\mathcal{B}} \\ \hat{\mathcal{B}}\mathcal{L}_{21} + \mathcal{L}_{12}^T b_0 & \Psi \end{bmatrix}, \tag{4.14}$$

where $\Psi \triangleq \hat{\mathcal{B}}\mathcal{L}_1 + \mathcal{L}_1^T \hat{\mathcal{B}}$, $\hat{\mathcal{B}} \triangleq \text{diag}\{b_1, b_2, \ldots, b_N\}$, $\mathcal{B}_0 = [b_0, 0, \ldots, 0]^T$ and it is easy to see that Ψ is positive.

Then, we can decompose the matrix $\check{\Psi}$ as follows:

$$\check{\Psi} = \begin{bmatrix} 0 & 0_{1\times N} \\ \hat{\mathcal{B}}\mathcal{L}_{21} + \mathcal{L}_{12}^T b_0 & \Psi \end{bmatrix}. \tag{4.15}$$

The local neighborhood tracking error is defined as follows:

$$\xi_i = \sum_{j=0}^{N} b_{ij}(\hat{r}_i - \hat{r}_j), i = 1, \ldots, N, \tag{4.16}$$

where

$$b_{ij} = \begin{cases} r_i a_{ij} + r_j a_{ji}, j \neq 0 \\ r_i a_{i0} + r_0 \bar{a}_i, j = 0. \end{cases} \tag{4.17}$$

The global dynamics is shown as follows:

$$\xi = (\Psi \otimes I_n)\varepsilon, \tag{4.18}$$

where $\xi = [\xi_1^T(t), \xi_2^T(t), \ldots, \xi_N^T(t)]^T, \varepsilon = [\varepsilon_1^T(t), \varepsilon_2^T(t), \ldots, \varepsilon_N^T(t)]^T, \varepsilon_i = \hat{r}_i - r_0$.

The measurement error $\epsilon_i(t)$ is defined as follows:

$$\epsilon_i = \hat{\mathcal{R}}_i - \hat{r}_i, i = 1, \ldots, N, \tag{4.19}$$

where $\hat{\mathcal{R}}_i = \hat{r}_i(t_k^i)$, t_k^i denotes the kth triggering instant of ith DG and determined by follows:

$$t_{k+1}^i = inf\{t > t_k^i : f_i(t) > 0\}. \tag{4.20}$$

The local neighborhood tracking error of the trigger state is defined as follows:

$$\hat{\xi}_i = \sum_{j=0}^{N} b_{ij}(\hat{\mathcal{R}}_i - \hat{\mathcal{R}}_j), i = 1, \ldots, N. \tag{4.21}$$

Auxiliary controllers u_i in (4.9) that rely on observer triggered states is designed as follows:

$$u_i = -\pi_i B^T P\hat{\xi}_i - c_i\gamma(B^T P\hat{\xi}_i), \tag{4.22}$$

where

$$\dot{\pi}_i = -k_{\pi_i}(\pi_i - 1) + \gamma_{\pi_i}\hat{\xi}_i^T PBB^T P\hat{\xi}_i, \tag{4.23}$$

$c_i, k_{\pi_i}, \gamma_{\pi_i}$ are positive constants, and

$$\gamma(\vartheta) = \begin{cases} \frac{\vartheta}{\|\vartheta\|}, & \|\vartheta\| > \chi_i \\ \frac{\vartheta}{\chi_i}, & \|\vartheta\| \le \chi_i, \end{cases} \tag{4.24}$$

with χ_i being small positive scalars, and $\|\cdot\|$ be 2-norm of matrix $\cdot$. Then,

$$\begin{aligned} \dot{\hat{r}}_i(t) =& A\hat{r}_i(t) - \pi_i\theta_i BB^T P\hat{\xi}_i - c_i\theta_i B\gamma(B^T P\hat{\xi}_i) + Bv_i^S \\ &+ B\varphi_i^T(\hat{r}_i)\hat{W}_i + LC\alpha_i, \end{aligned} \tag{4.25}$$

where

$$\dot{\hat{W}}_i = \gamma_{Wi}[\varphi_i(\hat{r}_i)y_i - k_{Wi}\hat{W}_i - 2\varphi_i(\hat{r}_i)\varphi_i^T(\hat{r}_i)\hat{W}_i], \tag{4.26}$$

and $\gamma_{Wi}, k_{Wi} > 0$. Denote $\alpha_i = \hat{r}_i - r_i$ as the observer error, and

$$\dot{\alpha}_i = (A - LC)\alpha_i + B(\varphi_i^T(\hat{r}_i)\hat{W}_i - \varphi_i^T(r_i)W_i - o_i). \tag{4.27}$$

It is not difficult to obtain that ε satisfy the following dynamics:

$$\begin{aligned}\dot{\varepsilon} =&(I_N \otimes A)\varepsilon - [\Pi\Theta \otimes BB^T P]\hat{\xi} - (\mathcal{C}\Theta \otimes B)\Gamma(\delta) \\ &+ (I_N \otimes B)v^S + (I_N \otimes B)\varphi_W + (I_N \otimes LC)\alpha, \end{aligned} \tag{4.28}$$

where $\Pi = \text{diag}\{\pi_1, \pi_2, \ldots, \pi_N\}$, $\Theta = \text{diag}\{\theta_1, \theta_2, \ldots, \theta_N\}$, $\hat{\xi} = [\hat{\xi}_1^T, \hat{\xi}_2^T, \ldots, \hat{\xi}_N^T]^T$, $\mathcal{C} = \text{diag}\{c_1, c_2, \ldots, c_N\}$, $\Gamma(\delta) = [\gamma(B^T P\hat{\xi}_1), \gamma(B^T P\hat{\xi}_2), \ldots, \gamma(B^T P\hat{\xi}_N)]^T$, $v^S = [v_1^S, v_2^S, \ldots, v_N^S]^T$, $\varphi_W = [\hat{W}_1^T\varphi_1(\hat{r}_1), \hat{W}_2^T\varphi_2(\hat{r}_2), \ldots, \hat{W}_N^T\varphi_N(\hat{r}_N)]$, $\alpha = [\alpha_1^T, \alpha_2^T, \ldots, \alpha_N^T]^T$, $\otimes$ be the Kronecker product.

Then, the sequence of triggered instants is determined by the following trigger mechanism:

$$f_i(t) = \epsilon_i^T PBB^T P\epsilon_i - \tau_i(t), \tag{4.29}$$

where $\tau_i(t) : [0, \infty) \to R^+$ is any positive scalar function that belong to $L_1[0, \infty)$ and be such that 1) $\inf_{t\in[t_a,t_b]} \tau_i(t) > 0$ for any finite $t_a, t_b > 0$ and 2) $\lim_{t\to\infty} \tau_i(t) = 0$.

Remark 4.3 Due to the inherent characteristics of the power system leading to the immeasurability of the local state information r_i, the proposed control protocol utilizes the local output y_i (v_{loadi}) and the trigger state of the neighboring observer for load regulation.

Remark 4.4 Although the controllers in the above methods of [5–7] are updated in an acyclic form, continuous access to adjacent states is still required in trigger error or threshold designs. This constant state transfer can still consume too much energy. A self-triggering implementation scheme based on sampled information is proposed in [6, 8, 9], transmitting only discrete-time information. Note, however, that to implement discrete-time measurements and discrete-time updates, the additional computational burden of each controller is usually used to "predict" future state information. In addition, the method will not be applicable in the case of agent dynamics with unknown terms. The proposed OBEDFT protocol consists of (4.17) and (4.20) since the trigger condition does not use the combined measurement error $e_i(t)$ but relies on consensus error $\hat{\xi}_i(t)$ and measurement error $\epsilon_i(t)$. Therefore, the neighbouring DGs do not need to communicate continuously.

Remark 4.5 Traditionally, fault factors are required to determine the controller gain. In contrast, in the proposed adaptive OBEDFT protocol, the parameter $\bar{\pi}$ associated with the fault factor is not in the control protocol. Thus, the protocol does not need information about the fault parameters. In addition, the boundary layer concept is introduced to approximate the nonsmooth function, thereby avoiding the undesirable chattering effect encountered by nonsmooth controllers [6]. Due to multiplicative and additive faults, the design process is very different from the general continuous controller in [10].

Theorem 4.1 *Assume that the communication topology is as described in Assumption 4.1. Under the controller (4.14) and the trigger condition (4.20), the consensus error converges exponentially to a finite state value, i.e., all DG's load voltages are restorable to the reference value v_{ref}. Moreover, $\Gamma > 0$, $P > 0$ in (4.20) and (4.14) are the solution to the below conditions:*

$$A^T\Gamma + \Gamma A + (\beta_2 + \beta_3)\Gamma BB^T\Gamma - C^TC + \beta_1 MM^T + 2I_2 < 0, \tag{4.30}$$

$$PA^T + AP - PBB^TP + 2I_2 = 0, \tag{4.31}$$

where $M = C^T + \Gamma B$, $L = \Gamma^{-1}C^T$, $\beta_1, \beta_2, \beta_3 > 0$.

Proof. Consider the following Lyapunov function candidate

$$V = V_1 + V_2, \tag{4.32}$$

with

$$V_1 = \sum_{i=1}^{N} \alpha_i^T \Gamma \alpha_i + \sum_{i=1}^{N} \frac{\tilde{W}_i^T \tilde{W}_i}{\gamma_{Wi}}, \tag{4.33}$$

and

$$V_2 = \varepsilon^T(\Psi \otimes P)\varepsilon + \sum_{i=1}^{N} \frac{(\pi_i - \bar{\pi})^2}{2\mu\gamma_{\pi_i}}, \tag{4.34}$$

where $\tilde{W}_i = \hat{W}_i - W_i$, and μ, $\bar{\pi}$ are positive constants, which are designed later. Because of $\pi_i(0) \geq 1$ and $\dot{\pi}_i \geq 0$, we can obtain that $\pi_i(t) \geq 1$ for any $t > 0$. Evidently, V is positive definite.

Part A: The time derivate of V_1 along the trajectory of (4.27) is obtained as follows:

$$\begin{aligned}\dot{V}_1 = &\sum_{i=1}^{N} \alpha_i^T[(A - LC)^T\Gamma + \Gamma(A - LC)]\alpha_i \\ &+ 2\sum_{i=1}^{N} \alpha_i^T \Gamma B(\varphi_i^T(\hat{r}_i)\hat{W}_i - \varphi_i^T(r_i)W_i - o_i) + 2\sum_{i=1}^{N} \frac{\tilde{W}_i^T \dot{\hat{W}}_i}{\gamma_{Wi}}.\end{aligned} \tag{4.35}$$

Define $\tilde{\varphi} = \varphi_i(\hat{r}_i) - \varphi_i(r_i)$. It is straightforward to show that

$$
\begin{aligned}
&2\sum_{i=1}^{N} \alpha_i^T \Gamma B(\varphi_i^T(\hat{r}_i)\hat{W}_i - \varphi_i^T(r_i)W_i - o_i) + 2\sum_{i=1}^{N} \frac{\tilde{W}_i^T \dot{\hat{W}}_i}{\gamma_{Wi}} \\
&\quad = 2\sum_{i=1}^{N} \alpha_i^T M \varphi_i^T(\hat{r}_i)\tilde{W}_i + \sum_{i=1}^{N} \alpha_i^T \Gamma B(\tilde{\varphi}_i^T W_i - o_i) \\
&\quad\quad + 2\sum_{i=1}^{N}(-k_{Wi}\tilde{W}_i^T \hat{W}_i - 2\tilde{W}_i^T \varphi_i(\hat{r}_i)\varphi_i^T(\hat{r}_i)\hat{W}_i).
\end{aligned} \tag{4.36}
$$

Using Young's inequality yields,

$$
\begin{aligned}
&2\sum_{i=1}^{N} \alpha_i^T M \varphi_i^T(\hat{r}_i)\tilde{W}_i + \sum_{i=1}^{N} \alpha_i^T \Gamma B(\tilde{\varphi}_i^T W_i - o_i) \\
&\leq \sum_{i=1}^{N} \beta_1 \alpha_i^T M M^T \alpha_i + \sum_{i=1}^{N} \beta_1^{-1} \tilde{W}_i^T \varphi_i(\hat{r}_i)\varphi_i^T(\hat{r}_i)\tilde{W}_i \\
&\quad + \sum_{i=1}^{N}(\beta_2 + \beta_3)\alpha_i^T \Gamma B B^T \Gamma \alpha_i + \sum_{i=1}^{N} \beta_2^{-1} W_i^T \tilde{\varphi}\tilde{\varphi}_i^T W_i + \sum_{i=1}^{N} \beta_3^{-1} o_i^T o_i,
\end{aligned} \tag{4.37}
$$

and

$$
\begin{aligned}
-2k_{Wi}\tilde{W}_i^T \hat{W}_i &= -2k_{Wi}\tilde{W}_i^T \tilde{W}_i + 2k_{Wi}\tilde{W}_i^T W_i \\
&\leq -k_{Wi}\tilde{W}_i^T \tilde{W}_i + k_{Wi} W_i^T W_i.
\end{aligned} \tag{4.38}
$$

Similarly,

$$
\begin{aligned}
&-4\sum_{i=1}^{N} \tilde{W}_i^T \varphi_i(\hat{r}_i)\varphi_i^T(\hat{r}_i)\hat{W}_i \\
&\leq -2\tilde{W}_i^T \varphi_i(\hat{r}_i)\varphi_i^T(\hat{r}_i)\tilde{W}_i + 2W_i^T \varphi_i(\hat{r}_i)\varphi_i^T(\hat{r}_i)W_i.
\end{aligned} \tag{4.39}
$$

Applying (4.36)–(4.39) to (4.35), because of lemma 4.2 and $\beta_1 \geq 1$, one can obtain that

$$\dot{V}_1 \leq \sum_{i=1}^{N} \alpha_i^T [A^T \Gamma + \Gamma A - C^T C + \beta_1 M M^T + (\beta_2 + \beta_3) \Gamma B B^T \Gamma] \alpha_i \\ + \sum_{i=1}^{N} [k_{Wi} + (4\beta_2^{-1} + 2) l_i] W_i^T W_i + \sum_{i=1}^{N} \beta_3^{-1} o_i^T o_i - k_{Wi} \tilde{W}_i^T \tilde{W}_i \\ - \tilde{W}_i^T \varphi_i(\hat{r}_i) \varphi_i^T(\hat{r}_i) \tilde{W}_i. \tag{4.40}$$

Part B: It from (4.25) follows that:

$$\dot{V}_2 = \varepsilon^T [\Psi \otimes (PA + A^T P)] \varepsilon - 2\varepsilon^T (\Psi \Pi \Theta \otimes PBB^T P) \hat{\xi} + 2\xi^T (I_N \otimes PB) v^S \\ - 2\xi^T (C\Theta \otimes PB) \Gamma(\delta) + 2 \sum_{i=1}^{N} \xi_i^T B \varphi_i^T(\hat{r}_i) \hat{W}_i + 2\varepsilon^T (\Psi \otimes PLC) \alpha \\ + \sum_{i=1}^{N} \frac{1}{\mu} (\pi_i - \bar{\pi}) \hat{\xi}_i^T PBB^T P \hat{\xi}_i - \sum_{i=1}^{N} \frac{k_{\pi_i}}{\mu \gamma_{\pi_i}} (\pi_i(t) - \bar{\pi})(\bar{\pi} - 1). \tag{4.41}$$

Denote $e_i = \hat{\xi}_i - \xi_i, i = 1, \ldots, N$, we have

$$-2\varepsilon^T (\Psi \Pi \Theta \otimes PBB^T P) \hat{\xi} \leq - \sum_{i=1}^{N} \pi_i \theta_i \xi_i^T PBB^T P \xi_i + \sum_{i=1}^{N} \pi_i \theta_i e_i^T PBB^T P e_i. \tag{4.42}$$

Moreover, one has

$$\sum_{i=1}^{N} \frac{1}{\mu} (\pi_i - \bar{\pi}) \hat{\xi}_i^T PBB^T P \hat{\xi}_i \\ \leq \frac{2}{\mu} \sum_{i=1}^{N} \pi_i \xi_i^T PBB^T P \xi_i + \frac{2}{\mu} \sum_{i=1}^{N} \pi_i e_i^T PBB^T P e_i \\ - \frac{1}{2\mu} (1 - \beta_4) \sum_{i=1}^{N} \bar{\pi} \xi_i^T PBB^T P \xi_i + \frac{1}{2\mu\beta_4} \sum_{i=1}^{N} \bar{\pi} \| B^T P \|^2 \| e_i \|^2 \\ - \frac{1}{2\mu} \sum_{i=1}^{N} \bar{\pi} \hat{\xi}_i^T PBB^T P \hat{\xi}_i. \tag{4.43}$$

According to the definition of θ_i and $\gamma(\chi)$, we get

Case 1: $\|B^T P\hat{\xi}_i\| > \chi_i, i = 1, 2, \ldots, N$

$$\begin{aligned}
&- 2\xi^T(\mathcal{C}\Theta \otimes PB)\Gamma(\delta) + 2\xi^T(I_N \otimes PB)v^S \\
&\le -2\sum_{i=1}^{N} c_i\theta_i \frac{\xi_i^T(t)PBB^TP\hat{\xi}_i}{\|B^TP\hat{\xi}_i\|} + 2\sum_{i=1}^{N} \varsigma_i\|B^TP\xi_i\| \\
&\le -2\sum_{i=1}^{N} c_i\theta_i \frac{(\hat{\xi}_i - e_i)^TPBB^TP\hat{\xi}_i}{\|B^TP\hat{\xi}_i\|} + 2\sum_{i=1}^{N} \varsigma_i\|B^TPe_i\| + 2\sum_{i=1}^{N} \varsigma_i\|B^TP\hat{\xi}_i\| \\
&\le 2\sum_{i=1}^{N}(c_i\theta_i + \varsigma_i)\|B^TPe_i\| + 2\sum_{i=1}^{N} \varsigma_i\|B^TP\hat{\xi}_i\| \\
&\le 2\sum_{i=1}^{N}(c_i\theta_i + \varsigma_i)\|B^TPe_i\| + 2\sum_{i=1}^{N} \frac{\varsigma_i}{\chi_i}\|B^TP\hat{\xi}_i\|^2.
\end{aligned} \tag{4.44}$$

Case 2: $\|B^T P\hat{\xi}_i\| \le \chi_i, i = 1, 2, \ldots, N$

$$\begin{aligned}
&- 2\xi^T(\mathcal{C}\Theta \otimes PB)\Gamma(\delta) + 2\xi^T(I_N \otimes PB)v^S \\
&\le 2\sum_{i=1}^{N} c_i\theta_i \frac{\|\xi_i^TPB\|\|B^TP\hat{\xi}_i\|}{\chi_i} + 2\sum_{i=1}^{N} \varsigma_i\|B^TP\xi_i\| \\
&\le 2\sum_{i=1}^{N}(c_i\theta_i + \varsigma_i)\|B^TP\xi_i\| \\
&\le 2\sum_{i=1}^{N}(c_i\theta_i + \varsigma_i)\|B^TPe_i\| + 2\sum_{i=1}^{N}(c_i\theta_i + \varsigma_i)\chi_i.
\end{aligned} \tag{4.45}$$

Case 3: Without loss of generality, assume that $\|B^T P\hat{\xi}_i\| > \chi_i, i = 1, 2, \ldots, q$ and $\|B^T P\hat{\xi}_i\| \le \chi_i, i = q+1, q+2, \ldots, N$, where $2 \le q \le N-1$, it can get

$$\begin{aligned}
&- 2\xi^T(\mathcal{C}\Theta \otimes PB)\Gamma(\delta) + 2\xi^T(I_N \otimes PB)v^S \\
&\le 2\sum_{i=1}^{N}(c_i\theta_i + \varsigma_i)\|B^TPe_i\| + 2\sum_{i=1}^{q} \frac{\varsigma_i}{\chi_i}\|B^TP\hat{\xi}_i\|^2 + 2\sum_{i=q+1}^{N}(c_i\theta_i + \varsigma_i)\chi_i.
\end{aligned} \tag{4.46}$$

From (4.44)–(4.46), we can obtain

$$\begin{aligned} &-2\xi^T(\mathcal{C}\Theta\otimes PB)\Gamma(\delta)+2\xi^T(I_N\otimes PB)v^S \\ &\le 2\sum_{i=1}^{N}(c_i\theta_i+\varsigma_i)\|B^TPe_i\|+2\sum_{i=1}^{N}(c_i\theta_i+\varsigma_i)\chi_i+2\sum_{i=1}^{N}\frac{\varsigma_i}{\chi_i}\|B^TP\hat{\xi}_i\|^2. \end{aligned} \tag{4.47}$$

Moreover,

$$\begin{aligned} &2\sum_{i=1}^{N}\xi_i^TPB\varphi_i^T(\hat{r}_i)\hat{W}_i \\ &=2\sum_{i=1}^{N}\xi_i^TPB\varphi_i^T(\hat{r}_i)\tilde{W}_i+2\sum_{i=1}^{N}\xi_i^TPB\varphi_i^T(\hat{r}_i)W_i \\ &\le 2\sum_{i=1}^{N}\xi_i^TPBB^TP\xi_i+\sum_{i=1}^{N}\tilde{W}_i^T\varphi_i(\hat{r}_i)\varphi_i^T(\hat{r}_i)\tilde{W}_i+\sum_{i=1}^{N}l_iW_i^TW_i. \end{aligned} \tag{4.48}$$

Because of $LCC^TL^T\le BB^T$, we have

$$2\varepsilon^T(\Psi\otimes PLC)\alpha\le\sum_{i=1}^{N}\xi_i^T(t)PBB^TP\xi_i(t)+\sum_{i=1}^{N}\alpha_i^T\alpha_i. \tag{4.49}$$

Applying (4.42), (4.43) and (4.47)–(4.49) to (4.41) we obtain

$$\begin{aligned} \dot{V}_2\le{}&\varepsilon^T[\Psi\otimes(PA+A^TP)]\varepsilon \\ &-\sum_{i=1}^{N}\Big[(\theta_i-\frac{2}{\mu})\pi_i+\frac{1}{2\mu}(1-\beta_4)\bar{\pi}-3\Big]\xi_i^TPBB^TP\xi_i \\ &+\sum_{i=1}^{N}(\pi_i\theta_i+\frac{2}{\mu}\pi_i+\frac{1}{2\mu\beta_4}\bar{\pi})e_i^T(t)PBB^TPe_i(t) \\ &+2\sum_{i=1}^{N}(c_i\theta_i+\varsigma_i)\|B^TPe_i\|-\frac{1}{2\mu}\sum_{i=1}^{N}\bar{\pi}\hat{\xi}_i^TPBB^TP\hat{\xi}_i \\ &+2\sum_{i=1}^{N}\frac{\varsigma_i}{\chi_i}\|B^TP\hat{\xi}_i\|^2+2\sum_{i=1}^{N}(c_i\theta_i+\varsigma_i)\chi_i-\sum_{i=1}^{N}\frac{k_{\pi_i}}{\mu\gamma_{\pi_i}}(\pi_i(t)-\bar{\pi})(\bar{\pi}-1) \\ &+\sum_{i=1}^{N}\tilde{W}_i^T\varphi_i(\hat{r}_i)\varphi_i^T(\hat{r}_i)\tilde{W}_i+\sum_{i=1}^{N}l_iW_i^TW_i+\sum_{i=1}^{N}\alpha_i^T\alpha_i. \end{aligned} \tag{4.50}$$

Moreover, for any positive constant β_5, we have

$$\begin{aligned}
&\sum_{i=1}^{N}(\pi_i\theta_i + \frac{2}{\mu}\pi_i + \frac{1}{2\mu\beta_4}\bar{\pi})e_i^T(t)PBB^TPe_i(t)\\
&\le \sum_{i=1}^{N}\Big[\frac{\beta_5}{2}(\pi_i - 1)^2 + \frac{1}{2\beta_5}(\theta_i + \frac{2}{\mu})^2\\
&+ (\theta_i + \frac{2}{\mu}) + \frac{1}{2\mu\beta_4}\bar{\pi}\Big]e_i^T(t)PBB^TPe_i(t). \qquad (4.51)
\end{aligned}$$

Because (4.16), (4.19) and (4.21), $e_i(t)$ can be transformed as $e_i(t) = \sum_{j=1}^{N} b_{ij}(\epsilon_i(t) - \epsilon_j(t))$, then

$$\begin{aligned}
&e_i^T(t)QBB^TQe_i(t)\\
&= \sum_{j=1}^{N} b_{ij}(\epsilon_i - \epsilon_j)^T QBB^TQ\sum_{j=1}^{N} b_{ij}(\epsilon_i - \epsilon_j)\\
&\le 2(\sum_{j=1}^{N} b_{ij})^2\epsilon_i^T QBB^TQ\epsilon_i + 2\sum_{j=1}^{N} b_{ij}\epsilon_j^T QBB^TQ\sum_{j=1}^{N} b_{ij}\epsilon_j\\
&\le 2(\sum_{j=1}^{N} b_{ij})^2\sum_{j=1}^{N}\epsilon_j^T QBB^TQ\epsilon_j, \qquad (4.52)
\end{aligned}$$

and

$$\begin{aligned}
\|B^TQe_i(t)\| &= \|\sum_{j=1}^{N} b_{ij}B^TQ\epsilon_i\| + \|\sum_{j=1}^{N} b_{ij}B^TQ\epsilon_j\|\\
&\le 2\sum_{j=1}^{N} b_{ij}\sum_{j=1}^{N}\|B^TQ\epsilon_j\|. \qquad (4.53)
\end{aligned}$$

Note that

$$\begin{aligned}
-(\pi_i - \bar{\pi})(\pi_i - 1) &= -(\pi_i - \bar{\pi})^2 - (\bar{\pi} - 1)(\pi_i - \bar{\pi})\\
&\le -\frac{1}{2}(\pi_i - \bar{\pi})^2 + \frac{1}{2}(\bar{\pi} - 1)^2, \qquad (4.54)
\end{aligned}$$

and

$$\begin{aligned}
-(\pi_i - \bar{\pi})(\pi_i - 1) &= -(\pi_i - 1)^2 + (\bar{\pi} - 1)(\pi_i - 1)\\
&\le -\frac{1}{2}(\pi_i - 1)^2 + \frac{1}{2}(\bar{\pi} - 1)^2. \qquad (4.55)
\end{aligned}$$

Combining (4.54) and (4.55), one can obtain

$$-\sum_{i=1}^{N}\frac{k_{\pi_i}}{\gamma_{\pi_i}\mu}(\pi_i-\bar{\pi})(\pi_i-1)$$
$$\leq\sum_{i=1}^{N}\frac{k_{\pi_i}}{2\gamma_{\pi_i}\mu}(\bar{\pi}-1)^2-\frac{k_{\pi_i}}{4\gamma_{\pi_i}\mu}(\pi_i-1)^2-\frac{k_{\pi_i}}{4\gamma_{\pi_i}\mu}(\pi_i-\bar{\pi})^2. \tag{4.56}$$

Since $0<\beta_4<1$, $\mu>2/\underline{\theta}_i$ and $\bar{\pi}_i\geq\max\{\mathcal{H}_i, 4\mu\varsigma_i/\chi_i\}$, $\mathcal{H}_i=2\mu\Big[\lambda_{\min}^{-1}(\Psi)+3+\underline{\theta}_i-2\mu^{-1}\Big]/(1-\beta_4)$ hold, trigger function (4.29) ensures that

$$\begin{aligned}\dot{V}_2\leq&-\varepsilon^T(\Psi\otimes I_2)\varepsilon+\sum_{i=1}^{N}\Big[\frac{1}{\beta_5}(\theta_i+\frac{2}{\mu})^2+2\left(\theta_i+\frac{2}{\mu}\right)+\frac{1}{\mu\beta_4}\bar{\pi}\Big]\left(\sum_{j=1}^{N}b_{ij}\right)^2\sum_{j=1}^{N}\tau_j\\&+2\sum_{i=1}^{N}(c_i\theta_i+\varsigma_i)\sum_{j=1}^{N}b_{ij}\sum_{j=1}^{N}(\tau_j)^{\frac{1}{2}}+2\sum_{i=1}^{N}(c_i\theta_i+\varsigma_i)\chi_i\\&+\sum_{i=1}^{N}\frac{k_{\pi_i}}{2\gamma_{\pi_i}\mu}(\bar{\pi}-1)^2-\Big[\frac{k_{\pi_i}}{4\gamma_{\pi_i}\mu}-\beta_5\sum_{j=1}^{N}b_{ij}\sum_{j=1}^{N}(\tau_j)^{\frac{1}{2}}\Big](\pi_i-1)^2\\&-\frac{k_{\pi_i}}{4\gamma_{\pi_i}\mu}(\pi_i-\bar{\pi})^2+\sum_{i=1}^{N}\tilde{W}_i^T\varphi_i(\hat{r}_i)\varphi_i^T(\hat{r}_i)\tilde{W}_i+\sum_{i=1}^{N}l_iW_i^TW_i+\sum_{i=1}^{N}\alpha_i^T\alpha_i.\end{aligned} \tag{4.57}$$

From (4.40) and (4.57), because of $\mu\geq\frac{k_{\pi_i}\beta_5(\sum_{j=1}^{N}b_{ij})^2\sum_{j=1}^{N}\tau_j(0)}{4\gamma_{\pi_i}}$ and (4.30) hold, we can obtain that

$$\begin{aligned}\dot{V}\leq&-\varepsilon^T(\Psi\otimes I_2)\varepsilon-\alpha^T(I_N\otimes I_2)\alpha\\&+\sum_{i=1}^{N}\left[\frac{1}{\beta_5}\left(\theta_i+\frac{2}{\mu}\right)^2+2\left(\theta_i+\frac{2}{\mu}\right)+\frac{1}{\mu\beta_4}\bar{\pi}\right]\left(\sum_{j=1}^{N}b_{ij}\right)^2\sum_{j=1}^{N}\tau_j\\&+2\sum_{i=1}^{N}(c_i\theta_i+\varsigma_i)\sum_{j=1}^{N}b_{ij}\sum_{j=1}^{N}(\tau_j)^{\frac{1}{2}}+2\sum_{i=1}^{N}(c_i\theta_i+\varsigma_i)\chi_i\\&+\sum_{i=1}^{N}\frac{k_{\pi_i}}{2\gamma_{\pi_i}\mu}(\bar{\pi}-1)^2-\frac{k_{\pi_i}}{4\gamma_{\pi_i}\mu}(\pi_i-\bar{\pi})^2\\&+\sum_{i=1}^{N}\left[k_{Wi}+(\beta_1^{-1}+4\beta_2^{-1}+1)l_i\right]W_i^TW_i+\sum_{i=1}^{N}\beta_3^{-1}o_i^To_i-\sum_{i=1}^{N}k_{Wi}\tilde{W}_i^T\tilde{W}_i.\end{aligned} \tag{4.58}$$

Note that (4.58) can be rewritten into

$$
\begin{aligned}
\dot{V} \leq & -\delta V-\varepsilon^T[\Psi \otimes (I_2-\delta P)]\varepsilon-\alpha^T[I_N \otimes (I_2-\delta \Gamma)]\alpha \\
& -\sum_{i=1}^N\left(\frac{k_{\pi_i}}{2}-\delta\right)\frac{(\pi_i(t)-\bar{\pi})^2}{2\mu\gamma_{\pi_i}}-\sum_{i=1}^N\left(k_{Wi}-\frac{\delta}{\gamma_{Wi}}\right)\tilde{W}_i^T\tilde{W}_i \\
& +\sum_{i=1}^N\left[\frac{1}{\beta_5}\left(\theta_i+\frac{2}{\mu}\right)^2+2\left(\theta_i+\frac{2}{\mu}\right)+\frac{1}{\mu\beta_4}\bar{\pi}\right]\left(\sum_{j=1}^N b_{ij}\right)^2\sum_{j=1}^N\tau_j \\
& +2\sum_{i=1}^N(c_i\theta_i+\varsigma_i)\sum_{j=1}^N b_{ij}\sum_{j=1}^N(\tau_j)^{\frac{1}{2}}+2\sum_{i=1}^N(c_i\theta_i+\varsigma_i)\chi_i \\
& +\sum_{i=1}^N\frac{k_{\pi_i}}{2\gamma_{\pi_i}\mu}(\bar{\pi}-1)^2+\sum_{i=1}^N\left[k_{Wi}+(\beta_1^{-1}+4\beta_2^{-1}+1)l_i\right]W_i^TW_i \\
& +\sum_{i=1}^N\beta_3^{-1}o_i^To_i.
\end{aligned}
\tag{4.59}
$$

Because of (4.30), (4.31) and $0<\delta\leq\min\{\frac{1}{\lambda_{\max}(P)},\frac{1}{\lambda_{\max}(\Gamma)},\frac{k_{\pi_i}}{2},k_{Wi}\gamma_{Wi}\}$, such that (4.54) satisfies the following inequality:

$$
\begin{aligned}
\dot{V} \leq & -\delta V+\sum_{i=1}^N\left[\frac{1}{\beta_5}\left(\theta_i+\frac{2}{\mu}\right)^2+2(\theta_i+\frac{2}{\mu})+\frac{1}{\mu\beta_4}\bar{\pi}\right]\left(\sum_{j=1}^N b_{ij}\right)^2\sum_{j=1}^N\tau_j \\
& +2\sum_{i=1}^N(c_i\theta_i+\varsigma_i)\sum_{j=1}^N b_{ij}\sum_{j=1}^N(\tau_j)^{\frac{1}{2}}+\ell.
\end{aligned}
\tag{4.60}
$$

V exponentially converges to a bounded set

$$
\mathcal{D}\triangleq\left\{\xi,c_i,\pi_i:V\leq\frac{1}{\delta}\ell\right\}, \tag{4.61}
$$

where

$$
\begin{aligned}
\ell \triangleq & \, 2\sum_{i=1}^N(c_i\theta_i+\varsigma_i)\chi_i+\sum_{i=1}^N\frac{k_{\pi_i}}{2\gamma_{\pi_i}\mu}(\bar{\pi}-1)^2 \\
& +\sum_{i=1}^N[k_{Wi}+(\beta_1^{-1}+4\beta_2^{-1}+1)l_i]W_i^TW_i+\sum_{i=1}^N\beta_3^{-1}o_i^To_i.
\end{aligned}
\tag{4.62}
$$

Therefore α_i, $\tilde{W}_i$, ξ, π_i and τ_i $i = 1, 2, \ldots, N$ are uniformly ultimately bounded. The proof is complete. ■

Theorem 4.2 *Under the controller (4.14) and the trigger condition (4.20), no Zeno phenomenon.*

Proof. The Dini derivative of $\|\epsilon_i\|$ is:

$$D^+\|\epsilon_i\| = \frac{d}{dt}\sqrt{\epsilon_i^T\epsilon_i} \leq \|\dot{\epsilon}_i\|. \tag{4.63}$$

for $t \in [t_k^i, t_{k+1}^i)$, we have

$$\begin{aligned}\dot{\epsilon}_i(t) = -\dot{\hat{r}}_i(t) = A\epsilon_i(t) + A\hat{\mathcal{R}}_i + \pi_i\theta_i BB^T P\hat{\xi}_i \\ + c_i\theta_i B\gamma(B^T P\hat{\xi}_i) - Bv_i^S + B\varphi_i^T(\hat{r}_i)\hat{W}_i - LC\alpha_i.\end{aligned} \tag{4.64}$$

Denote

$$\begin{aligned}\sigma_i = \max_{t\in[t_k^i, t_{k+1}^i]} \|A\hat{\mathcal{R}}_i\| + \|\pi_i\theta_i BB^T P\hat{\xi}_i\| + \|c_i\theta_i B\gamma(B^T P\hat{\xi}_i)\| \\ + \|Bv_i^S\| + \|B\varphi_i^T(\hat{r}_i)\hat{W}_i\| + \|LC\alpha_i\|.\end{aligned} \tag{4.65}$$

Then,

$$\frac{d\|\epsilon_i(t)\|}{dt} \leq \|A\|\|\epsilon_i(t)\| + \sigma_i. \tag{4.66}$$

By Comparison Principle, $\|\epsilon_i(t)\|$ satisfies:

$$\|\epsilon_i(t)\| = \frac{\sigma_i}{\|A\|}(e^{\|A\|(t-t_k^i)} - 1). \tag{4.67}$$

Taking notice of that

$$\|\epsilon_i\| \leq \frac{\sqrt{\tau_i}}{\|B^T P\|} \triangleq \epsilon_i^*. \tag{4.68}$$

Then, from the following inequality:

$$\frac{\sigma_i}{\|A\|}(e^{\|A\|\tau_k^i} - 1) \geq \epsilon_i^*, \tag{4.69}$$

we obtain a lower bound τ_k^i of $t_{k+1}^i - t_k^i$.

Based on (4.67), (4.68) and (4.55), we have

$$t^i_{k+1} - t^i_k \geq \tau^i_k \geq \frac{1}{\|A\|} ln(1 + \frac{\|A\|}{\sigma_i} \epsilon^*_i). \tag{4.70}$$

Then, no Zeno phenomenon.

■

4.3 Simulation Results

In this section, to demonstrate the feasibility and effectiveness of the proposed OBEDFT protocol for regulating the load voltage in the presence of PLOE faults and bias faults of actuators, we simulate the MG shown in Fig. 4.2, in several scenarios in the MATLAB/Simulink software environment. The isolated MG is tested with a combination of four inverter-based DGs. Table 4.1 provides the parameters of DGs, lines, and loads in Fig. 4.2. It is assumed that the DGs communicate with each other through the bidirectional asymmetric topology shown in Fig. 4.2. DG#1 and DG#4 can receive information from v_{ref}.

4.3.1 *Case 1: Voltage Regulation Effectiveness*

We tested the effectiveness of the OBEDFT load voltage regulation strategy.

Solving (4.30) and (4.31) gives a solution

$$P = \begin{bmatrix} 1.2100 \; 0.7321 \\ 0.7321 \; 2.0958 \end{bmatrix},$$

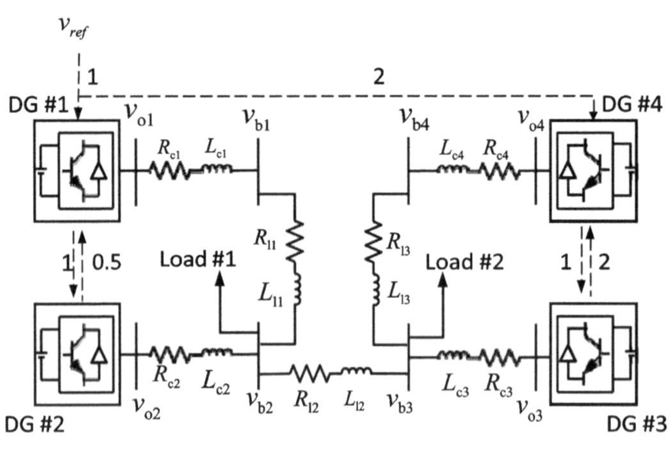

Fig. 4.2 The example MG test system and considered communication structure

Table 4.1 Specifications of the MG test system

DGs	DG#1 & DG#2 & DG#3 & DG#4	
	m_P	1.5×10^{-5}
	n_Q	2×10^{-4}
Lines	$R_{l1} = R_{l3} = 1e - 4\Omega$	$R_{l2} = 1e - 4\,\Omega$
	$L_{l1} = L_{l3} = 0.318\,\text{emH}$	$L_{l2} = 1.847\,\text{mH}$
RL loads	Load #1 & Load #2	
	P(per phase)	Q(per phase)
	50 kW	120 kVAr

From Fig. 4.2, we can derive its Laplace matrix as

$$\mathcal{L} = \begin{bmatrix} 0 & 0 & 0 & 0 & 0 \\ -1 & 1.5 & -0.5 & 0 & 0 \\ 0 & -1 & 1 & 0 & 0 \\ 0 & 0 & 0 & 1 & -1 \\ -2 & 0 & 0 & -2 & 4 \end{bmatrix}.$$

We choose the control parameters of the OBEDFT protocol as $\tau_1 = 16$, $\tau_1 = \tau_2 = \tau_3 = 14.5$, $k_{\pi_i} = 1$, $\gamma_i = 0.01$, $\gamma_{\pi_i} = 5 \times 10^6$, $c_i = 10^6$, $\chi_i = 1$, $\gamma_{Wi} = 1$, $k_{Wi} = 30$, $i = 1, 2, \ldots, 4$. Let $\pi_i(0.8) = 1$ and $\hat{W}_i(0.8) = 1$ be the initial value of $\pi_i(t)$ and $\hat{W}_i(t)$, $i = 1, 2, \ldots, N$, respectively.

We choose the PLOE faults and bias faults as $\theta_1 = 0.4\cos(5t - 3) + 0.7$, $\theta_2 = 0.1\cos(3t - 3) + 0.1$, $\theta_3 = 2$, $\theta_4 = 2\cos(5t - 2) + 0.1$, $v_1^S = 4\cos(4t - 1) + 3$, $v_2^S = 3\sin(0.5t - 4) + 5$, $v_3^S = 0.1\cos(t - 8) + 0.1$ and $v_4^S = 2\sin(3t - 2) + 0.66$.

To verify the effectiveness of the OBEDFT protocol, we relied on only one control at the beginning, and the load voltages deviated from the nominal value, as shown in Figs. 4.3 and 4.4. At t = 0.8 s, the OBEDFT load voltage secondary control cuts in, and the load voltages return to the reference value. At $t = 1.3$ s, load #1 cuts off 50%, and the load voltages fluctuates somewhat, but a short time later, the load voltages return to the reference value. At t = 1.8 s, 50% of load #1 is restored to its original value. Moreover, the trigger moments of the four DGs are shown in Fig. 4.5, which also illustrates that this indicates no Zeno phenomenon.

Therefore, the OBEDFT load voltage regulation strategy proposed in this chapter is proven effective, and the same applies to the load transformation case.

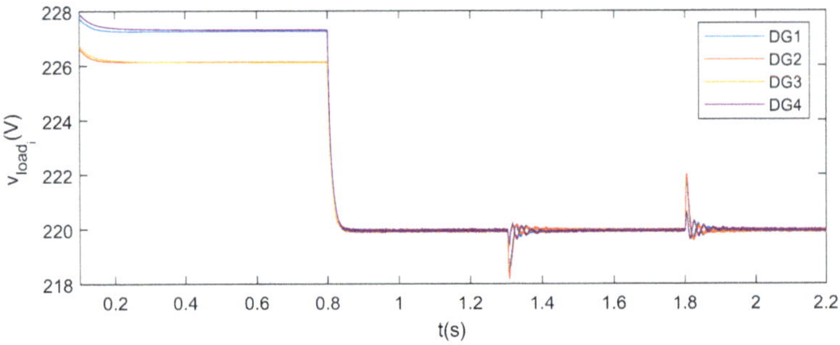

Fig. 4.3 Load volatge

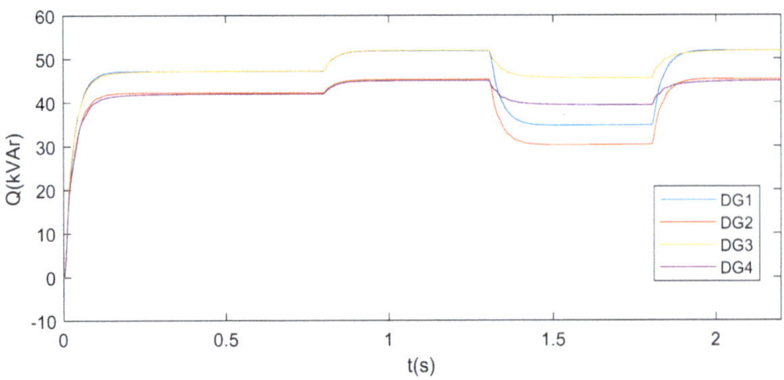

Fig. 4.4 Reactive power

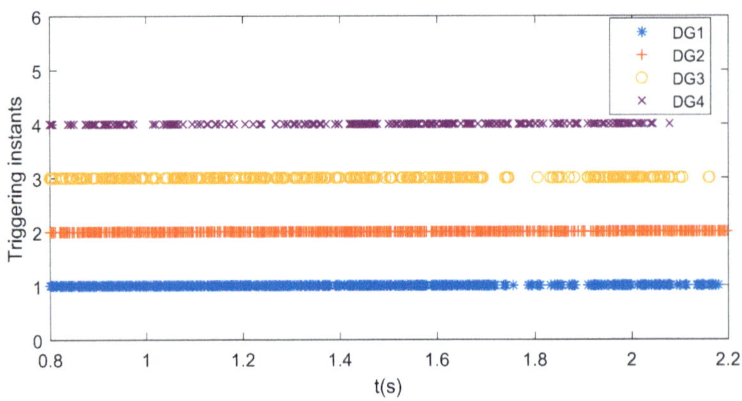

Fig. 4.5 Triggering instans

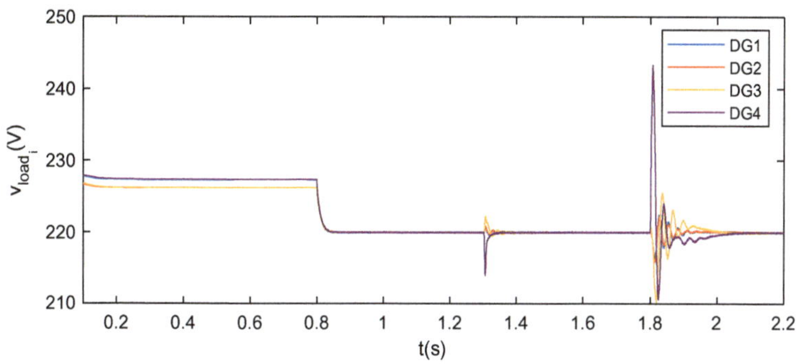

Fig. 4.6 Load volatge

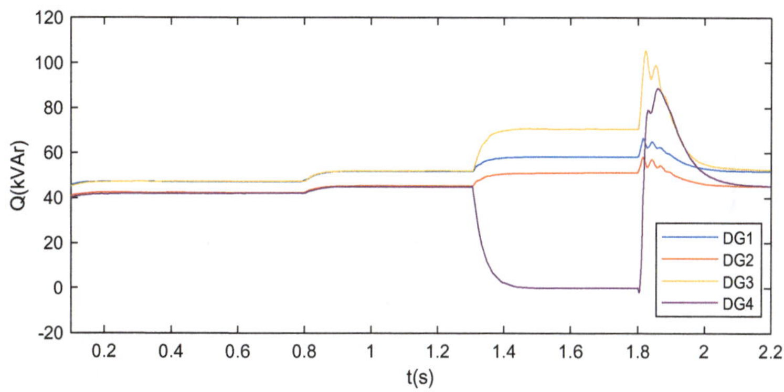

Fig. 4.7 Reactive power

4.3.2 Case 2: PIPO Capability

First, we change the fault parameters to test the effectiveness of the OBEDFT load voltage regulation strategy for different faults. Next, we test the design's PIPO capability.

$\tau_1 = \tau_2 = \tau_3 = \tau_4 = 13$ and other control parameters are the same as case 1. The PLOE faults and bias faults in the DGs are changed as follows: $\theta_1 = 1$, $\theta_2 = 0.7\cos(3t-3)+1$, $\theta_3 = 0.1$, $\theta_4 = 0.1\sin(5t-2)+0.4$, $v_1^S = 2\sin(3t-3)+0.2$, $v_2^S = 1.4\sin(5t-7)+0.1$, $v_3^S = 5\cos(2t-10)+0.4$, $v_4^S = 3\cos(3t-2)+0.1$. The secondary control is activated at $t = 0.8$ s. DG#4 was disconnected at $t = 1.3$ s and reconnected at $t = 1.8$ s.

Figures 4.6 and 4.7 shows that the proposed OBEDFT load voltage regulation strategy still performs well after the fault parameters are changed, even if DG#4 goes through the disconnection and reconnection process. And Fig. 4.8 presents the trigger moments of the four DGs.

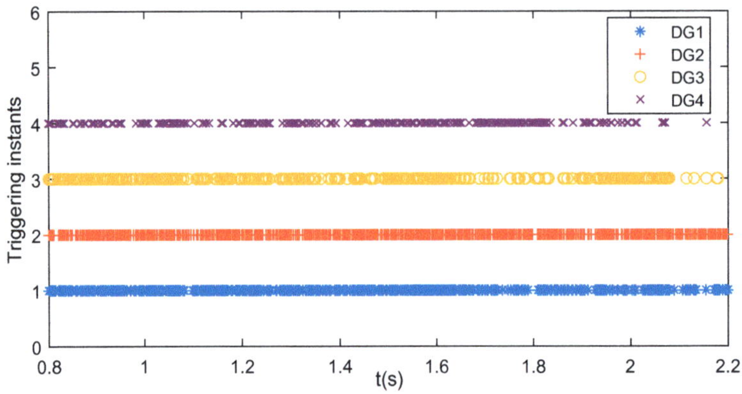

Fig. 4.8 Triggering instans

4.4 Conclusions

The load voltage regulation problem has been first transformed into a distributed output feedback tracking problem for linear MASs with nonlinear functions. The unavailability of full-state measurements was considered. The necessary preprocessing has been provided to implement the proposed OBEDFT control strategy for load voltage recovery to the reference value. The protocol compensates for PLOE and deviation faults of actuators with heterogeneity and time variability. The method does not require continuous communication between neighboring DGs'. Adaptive and boundary layer techniques have been introduced to avoid the bounds of fault signals and undesirable chattering effects encountered by non-smooth controllers. It is suitable for the case where the bidirectional communication topology may have asymmetric weights.

References

1. Bidram, A., Davoudi, A., Lewis, F.L., Guerrero, J.M.: Distributed cooperative secondary control of microgrids using feedback linearization. IEEE Trans. Power Syst. **28**(3), 3462–3470 (2013)
2. Wang, L.X., Mendel, J.M.: Fuzzy basis functions, universal approximation, and orthogonal least-squares learning. IEEE Trans. Neural Netw. **3**(5), 807–814 (1992)
3. Deng, C., Yang, G.H.: Distributed adaptive fuzzy control for nonlinear multi-agent systems under directed graphs. IEEE Trans. Fuzzy Syst. **26**(99), 1356–1366 (2018)
4. Xu, Y., Wu, Z.G.: Distributed adaptive event-triggered fault-tolerant synchronization for multiagent systems. IEEE Trans. Ind. Electron. **68**(2), 1537–1547 (2021)
5. Guo, B., Chen, Y.: Event-triggered robust adaptive sliding mode fault-tolerant control for nonlinear systems. IEEE Trans. Ind Inf **16**(11), 6982–6992 (2020)

6. Guo, X.G., Liu, P.M., Wang, J.L., Ahn, C.K.: Event-triggered adaptive fault-tolerant pinning control for cluster consensus of heterogeneous nonlinear multi-agent systems under aperiodic dos attacks. IEEE Trans. Netw. Sci. Eng. **8**(2), 1941–1956 (2021)
7. Li, Z., Yan, J., Yu, W., Qiu, J.: Event-triggered control for a class of nonlinear multiagent systems with directed graph. IEEE Trans. Syst. Man Cybern. Syst. **51**(11), 6986–6993 (2021)
8. Chen, J., Chen, B. , Zeng, Z.: Adaptive dynamic event-triggered fault-tolerant consensus for nonlinear multiagent systems with directed/undirected networks. IEEE Trans. Cybern. (2022)
9. Sun, J., Tan, Z., Liu, S. , Zhang, H., Chuo, W.: Fully distributed event-driven coordination with actuator faults. IEEE Trans. Cybern. (2022)
10. Zhai, M. ,Sun, Q. , Wang, B., Liu, Z., Zhang, H.: Cooperative fault-estimation-based event-triggered fault-tolerant voltage restoration in islanded AC microgrids. IEEE Trans. Autom. Sci. Eng. (2022)

Chapter 5
Secondary Load Voltage and Frequency Fault-Tolerant Regulation Control via Hierarchical Mechanism

Abstract This chapter presents a distributed containment-based load voltage and consensus-based frequency fault-tolerant secondary control strategy. The strategy considers distributed generators (DGs) that communicate through a directed network whose fault models include both actuators' partial loss of effectiveness (PLOE) and deviation faults. First, we introduce the idea of containment control in the load voltage secondary control strategy. Unlike the traditional consensus-based output voltage secondary control strategy, this strategy limits the load voltage on each bus to a reasonable range while having a voltage difference between the loads to allow power flow. Next, we use a hierarchical mechanism to decouple the distributed secondary control problem into a cooperative control problem for virtual systems and a decentralized control problem for the existing system. Note that the strategy avoids using global information about the directed communication network and quantitative information about the size, frequency, and type of actuator faults. In addition, we design a consensus-based frequency regulation and power-sharing control strategy. Finally, simulation results verify the effectiveness of the process.

Keywords Containment control · Voltage regulation · Fault-tolerant control · Adaptive control

5.1 Introduction

In this chapter, we study the AC MG containment-based load voltage regulation and consensus-based frequency fault-tolerant secondary control strategy. The voltage regulation at the load side is considered. And if multiple load voltages are controlled at the reference value simultaneously, it is impossible to achieve power flow between buses. For this reason, the control objective containing the control is introduced into the secondary control strategy to balance the conflicting goals of voltage regulation and power flow. This control strategy enables the power flow by controlling the load voltages within a reasonable range in the presence of voltage differences between the buses. A fault-tolerant hierarchical control method based on a virtual system is proposed for the complex decoupling problem between time-varying PLOE faults

J. Sun et al., *Building a Low-Carbon Future,* Studies in Systems, Decision and Control 506, https://doi.org/10.1007/978-3-031-43498-3_5

and adaptive controller gains under a directed communication network. The upper layer is based on the virtual system of neighborhood state information so that all virtual systems track into the convex packet formed by the leaders. The lower layer is designed with a decentralized adaptive tracking controller to achieve synchronization between the virtual system and the actual system. This control strategy does not require quantitative information about the magnitude, frequency, and type of PLOE and bias faults in the actuators. It only needs to ensure the boundedness of the fault signal. The proposed control strategy relies only on the relevant information of the neighboring virtual reference systems and not on the global information.

Notation: Let $\|A\|$ be 2-norm of matrix A, col$\{a_1, \ldots, a_N\}$ be a column vector and $\otimes$ be the Kronecker product, $\lambda_{\max}(B)$ be maximum eigenvalues of matrix B, diag$\{b_1, \ldots, b_N\}$ be a diagonal matrix, respectively.

5.2 Problem Statement and Preliminaries

The chapter focuses on the regulation of load voltage, frequency, and power in a MG system consisting of N DGs. Figure 5.1 illustrates the structure of each DG, which includes droop control, inner-loop voltage-current control, LC filter, secondary controller, and line model. The secondary control consists of containment-based voltage control and consensus-based frequency and power regulation control.

5.2.1 Preliminaries on Graph Theory

In an MG, DGs communicate in a directed network. We describe the communication topology of DGs using a directed graph $\mathbb{G}(\mathcal{V}, \mathcal{E}, \mathcal{A})$, where $\mathcal{V} = \vec{v}_1, \vec{v}_2, \ldots, \vec{v}_N\}$ is a non-empty finite set of nodes, $\mathcal{E} \subset \mathcal{V} \times \mathcal{V}$ is the set of edges, and $\mathcal{A}$ is the adjacency matrix. If $(\vec{v}_i, \vec{v}_j) \in \mathcal{T}$, then $a_{ij} = 1$; otherwise, $a_{ij} = 0$. The Laplacian matrix $\mathfrak{L}$ satisfies $\mathfrak{L}_{ii} = \sum_{j=0}^{N} a_{ij}$ and $\mathfrak{L}_{ij} = -a_{ij}$, where $i \neq j$. Graph $\mathbb{G}$ contains a directed spanning tree if it has a vertex that can communicate with all other vertices via directed paths. If graph $\mathbb{G}$ has a vertex that can communicate to all other vertices through a directed path, we say $\mathbb{G}$ contains a directed spanning tree.

5.2.2 Modeling Framework

The ith DG uses the following P/ω and Q/V droop control to ensure voltage and frequency stability:

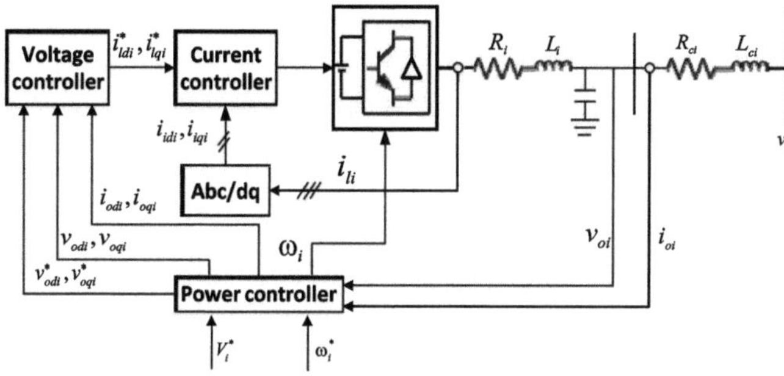

Fig. 5.1 Block diagram of a multi-MG system and the proposed containment-based secondary voltage regulation strategy

$$
\begin{aligned}
\omega_i &= \omega_i^* - m_i P_i, \\
v_{mag,i}^* &= V_i^* - k_i Q_i,
\end{aligned}
\tag{5.1}
$$

where $v_{mag,i}^*$ is the voltage amplitude at the inverter terminal, ω_i is the angular frequency of the voltage; ω_i^* and V_i^* indicate the inverter at the primary control frequency and voltage reference value, respectively; m_i and k_i denote the droop coefficients, respectively; P_i and Q_i are the active and reactive power, respectively.

In a rotating coordinate system based on its output voltage vector, the reactive droop equation yields the voltage as the d-axis reference voltage, and the q-axis reference voltage is $v_{oqi} = 0$. Then,

$$
v_{odi} = V_i^* - k_i Q_i. \tag{5.2}
$$

The load voltage can be written as

$$
v_{bdi} = v_{odi} + v_{li}, \tag{5.3}
$$

where v_{li} represents the voltage between the load and the inverter output. The essence of the load voltage $v_{bdi}(t)$ regulation is achieved by adjusting V_i^*.

Differentiating (5.1), one obtains

$$
\begin{aligned}
\dot{\omega}_i &= \dot{\omega}_i^* - m_i \dot{P}_i = u_{fi}, \\
\dot{v}_{odi} &= \dot{V}_i^* - k_i \dot{Q}_i = u_{vi},
\end{aligned}
\tag{5.4}
$$

where $u_{fi}(t)$ and $u_{vi}(t)$ are the frequency and voltage controllers. Then,

$$
\dot{v}_{bdi} = \dot{v}_{odi} + \dot{v}_{li} = u_{vi} + \dot{v}_{li}. \tag{5.5}
$$

Furthermore, let

$$m_i \dot{P}_i = u_{pi}, \tag{5.6}$$

where $u_{pi}(t)$ is an active power controller. Then,

$$\begin{aligned} \omega_i^* &= \int (u_{fi} + u_{pi})dt, \\ V_i^* &= \int (k_i \dot{Q}_i + u_{vi})dt. \end{aligned} \tag{5.7}$$

Hence, our goal is to design $u_{vi}(t)$, $u_{fi}(t)$, and $u_{pi}(t)$, so that the load voltage, frequency and active power can be regulated efficiently.

Remark 5.1 Unlike the traditional output voltage v_{odi} regulation problem in [1–9], in this chapter, we address the load voltage v_{bdi} regulation problem for AC MGs. The load voltage regulation problem is transformed into a distributed tracking problem with linear MASs, where the system contains nonlinear dynamics.

5.2.3 Fault Model

Signal faults can be modeled as:

$$\begin{aligned} u_{vi}^F &= \mathcal{Q}_{vi} u_{vi} + \mathcal{P}_{vi}, \\ u_{\omega i}^F &= \mathcal{Q}_{\omega i} u_{\omega i} + \mathcal{P}_{\omega i}, \\ u_{Pi}^F &= \mathcal{Q}_{pi} u_{pi} + \mathcal{P}_{pi}, \end{aligned} \tag{5.8}$$

where $\mathcal{Q}_{vi}, \mathcal{Q}_{\omega i}, \mathcal{Q}_{Pi}$ are PLOE fault factors of the ith DG's actuator of voltage, frequency and active power. $\mathcal{P}_{vi}, \mathcal{P}_{\omega i}, \mathcal{P}_{Pi}$ are biased fault factors of the ith DG's actuator of voltage, frequency and active power.

Assumption 5.1 $\mathcal{Q}_{vi}(t)$ with $0 < \underline{\mathcal{Q}}_{vi} \leq \mathcal{Q}_{vi}(t) \leq \overline{\mathcal{Q}}_{vi}$, $\mathcal{Q}_{\omega i}(t)$ with $0 < \underline{\mathcal{Q}}_{\omega i} \leq \mathcal{Q}_{\omega i}(t) \leq \overline{\mathcal{Q}}_{\omega i}$, $\mathcal{Q}_{Pi}(t)$ with $0 < \underline{\mathcal{Q}}_{Pi} \leq \mathcal{Q}_{Pi}(t) \leq \overline{\mathcal{Q}}_{Pi}$ are time-varying. $\underline{\mathcal{Q}}_{vi}$, $\overline{\mathcal{Q}}_{vi}$, $\underline{\mathcal{Q}}_{\omega i}$, $\overline{\mathcal{Q}}_{\omega i}$, $\underline{\mathcal{Q}}_{Pi}$ and $\overline{\mathcal{Q}}_{Pi}$ are unknown positive constants.

Assumption 5.2 $\|\mathcal{P}_{vi}\| \leq \varsigma_{vi}$, $\|\mathcal{P}_{\omega i}\| \leq \varsigma_{\omega i}$, $\|\mathcal{P}_{Pi}\| \leq \varsigma_{Pi}$ with $\varsigma_{vi}, \varsigma_{\omega i}, \varsigma_{Pi} > 0$ being unknown constants.

5.3 Design of Secondary Control Strategy

5.3.1 The Containment-Based Load Voltage Regulation Control

The containment-based voltage control bound voltages within a reasonable range. The virtual leader is defined as $\upsilon_k = \text{col}\{\upsilon_{N+1}, \upsilon_{N+2}\}$, $\upsilon_{N+1}, \upsilon_{N+2}$ represent the upper and lower boundaries of the specified voltage regulation, respectively. Moreover,

$$\dot{\upsilon}_k = 0, k = N + 1, N + 2. \tag{5.9}$$

Assumption 5.3 For each DGs, the virtual leaders' dynamics have paths to all DGs. Then, the Laplacian matrix $\mathfrak{L}$ of the communication network $\mathbb{G}$ among the virtual leaders and the DGs is defined as

$$\mathfrak{L} = \begin{bmatrix} 0 & 0_{2\times N} \\ \mathfrak{L}_2 & \mathfrak{L}_1 \end{bmatrix}, \tag{5.10}$$

where $\mathfrak{L}_2 \in \mathbb{R}^{N\times 2}$, $\mathfrak{L}_1 \in \mathbb{R}^{N\times N}$. $\mathfrak{L}_1$ is a nonsingular M-matrix.

Lemma 5.1 *[10] The matched uncertainty $f_i(\upsilon_{bdi})$ can be linearly parameterized by an neural network as*

$$f_i(\upsilon_{bdi}) = W_{\upsilon i}^T \varphi_{\upsilon i}(\upsilon_{bdi}) + \varepsilon_i, \tag{5.11}$$

where $W_{\upsilon i} \in \mathbb{R}^{s\times 1}$ is an unknown constant ideal weight matrix that satisfies $\|W_{\upsilon i}\| \leq \check{W}_{\upsilon i}$ with $\check{W}_i$ a positive constant, $\varphi_{\upsilon i}(\cdot) : \mathbb{R}^n \to \mathbb{R}^s$ is a known vector function of the form $\varphi_{\upsilon i}(\upsilon_{bdi}) = [\varphi_{\upsilon i1}(\upsilon_{bdi}), \varphi_{\upsilon i2}(\upsilon_{bdi}), \ldots, \varphi_{\upsilon is}(\upsilon_{bdi})]^T$ and satisfies $\|\varphi_{\upsilon i}\| \leq \check{\varphi}_{\upsilon i}$, $\check{\varphi}_{\upsilon i}$ is a positive constant, ε_i is the approximation error satisfying $\|\varepsilon_{\upsilon i}\| \leq \check{\varepsilon}_{\upsilon i}$, $\check{\varepsilon}_{\upsilon i}$ is a positive constant.

By Lemma 5.1, we obtain

$$\dot{\upsilon}_{li} = W_{\upsilon i}^T \varphi_{\upsilon i}(\upsilon_{bdi}) + \varepsilon_i. \tag{5.12}$$

From (5.5), rewrite υ_{bdi} as follows:

$$\dot{\upsilon}_{bdi} = \mathcal{Q}_{\upsilon i} u_{\upsilon i} + \mathcal{P}_{\upsilon i} + \dot{\upsilon}_{li}, i = 1, \ldots, N. \tag{5.13}$$

Let the controller $u_{\upsilon i}$ in (5.13) be

$$u_{\upsilon i} = -(\kappa_{\upsilon i} + \ell_{\upsilon i})\vartheta_{\upsilon i}(t) + c_{\upsilon i}\eta(\varpi_{\upsilon i}(t)), \tag{5.14}$$

where

$$\dot{\kappa}_{vi} = -\nu_{\kappa_{vi}}(\kappa_{vi} - 1) + \vartheta_{vi}^2(t), \tag{5.15}$$

$$\ell_{vi} = \vartheta_{vi}^2(t), \tag{5.16}$$

$$\dot{c}_{vi} = -k_{c_{vi}}c_{vi} + \gamma_{c_{vi}}\|\varpi_{vi}\|, \tag{5.17}$$

and

$$\eta(*) = \begin{cases} \frac{*}{\|*\|} + c_{vi}*, & if \ \|*\| > \chi_{vi} \\ \frac{*}{\chi_{vi}} + c_{vi}*, & if \ \|*\| \leq \chi_{vi}, \end{cases} \tag{5.18}$$

where χ_{vi} is a positive constant. $\vartheta_{vi}(t)$, $\varpi_{vi}(t)$ will be designed later.

Then, we can obtain

$$\dot{\upsilon}_{bdi} = -\mathcal{Q}_{vi}(\kappa_{vi} + \ell_{vi})\vartheta_{vi}(t) + \mathcal{Q}_{vi}c_{vi}\eta(\varpi_{vi}(t)) + \mathcal{P}_{vi} + \dot{\upsilon}_{li}. \tag{5.19}$$

The virtual system $\hat{\upsilon}_{bdi}$ to estimate υ_{bdi}, $i = 1, 2, \ldots, N$ as follows:

$$\dot{\hat{\upsilon}}_{bdi} = \hat{u}_{vi} + \hat{W}_{vi}^T\varphi_{vi}(\hat{\upsilon}_{bdi}) - l_{vi}(\hat{\upsilon}_{bdi} - \upsilon_{bdi}), \tag{5.20}$$

The controller of the virtual system can be designed as

$$\hat{u}_{vi} = -(\kappa_{vi} + \ell_{vi})\vartheta_{vi}(t). \tag{5.21}$$

Moreover

$$\dot{\hat{W}}_{vi} = -\gamma_{wvi}[j_{vi}\varphi_{vi}(\hat{\upsilon}_{bdi})\varpi_{vi} + k_{vi}\hat{W}_{vi}], \tag{5.22}$$

where γ_{wvi}, and k_{vi} are positive constants, and $j_{vi} \geq 1$.

Define $\varpi_{vi} = \hat{\upsilon}_{bdi} - \upsilon_{bdi}$, then

$$\begin{aligned} \dot{\varpi}_{vi} = &-(1 - \mathcal{Q}_{vi})(\kappa_{vi} + \ell_{vi})\vartheta_{vi}(t) - c_{vi}\mathcal{Q}_{vi}\eta(\varpi_{vi}(t)) - \mathcal{P}_{vi} \\ &- l_{vi}\varpi_{vi} + \hat{W}_{vi}^T\varphi_{vi}(\hat{\upsilon}_{bdi}) - W_{vi}^T\varphi_{vi}(\upsilon_{bdi}) - o_{vi}. \end{aligned} \tag{5.23}$$

Define the global error of the virtual system as:

$$\vartheta_{vi}(t) = \sum_{j=1}^{N} a_{ij}(\hat{v}_{bdi} - \hat{v}_{bdj}) + \sum_{k=N+1}^{N+2} g_{vi}(\hat{v}_{bdi} - v_k), i = 1, \ldots, N. \tag{5.24}$$

Moreover, define contain error

$$\varsigma = \hat{v}_{bd} - \bar{v}_{bd} = \hat{v}_{bd} + \mathfrak{L}_1^{-1}\mathfrak{L}_2 v_k, \tag{5.25}$$

where $\varsigma = \text{col}\{\varsigma_1(t), \varsigma_2(t), \ldots, \varsigma_N(t)\}$, $\hat{v}_{bd} = \text{col}\{\hat{v}_{bd1}, \hat{v}_{bd2}, \ldots, \hat{v}_{bdN}\}$.

The global vector ϑ_v is represented as

$$\vartheta_v = \mathfrak{L}_1\hat{v}_{bd} + \mathfrak{L}_2 v_k. \tag{5.26}$$

Remark 5.2 If the load voltage v_{bdi}, $i = 1, 2, \ldots, N$ are controlled at its nominal value, the power cannot flow between DGs. Therefore, this chapter introduces the control objective of containment control. On this basis, a secondary voltage regulation strategy based on containment control is proposed. The strategy limits each load voltage v_{bdi} to a reasonable range while having a voltage difference between DGs to allow power flow.

Moreover,

$$\dot{\vartheta}_v = -\mathfrak{L}_1(\kappa_v + \ell_v)\vartheta_v(t) + \mathfrak{L}_1\hat{W}_v^T\varphi(\hat{v}_{bd}) - \mathfrak{L}_1 L\varpi_v, \tag{5.27}$$

where $\kappa_v = \text{diag}\{\kappa_{v1}, \kappa_{v2}, \ldots, \kappa_{vN}\}$, $\ell_v = \text{diag}\{\ell_{v1}, \ell_{v2}, \ldots, \ell_{vN}\}$, $\vartheta_v = \text{diag}\{\vartheta_{v1}, \vartheta_{v2}, \ldots, \vartheta_{vN}\}$, $\hat{W}_v = \text{col}\{\hat{W}_{v1}, \hat{W}_{v2}, \ldots, \hat{W}_{vN}\}$, $\varphi(\hat{v}_{bd}) = \text{diag}\{\varphi_1(\hat{v}_{bd1}), \varphi_2(\hat{v}_{bd2}), \ldots, \varphi_N(\hat{v}_{bdN})\}$, $L = \text{diag}\{l_{v1}, l_{v2}, \ldots, l_{vN}\}$, $\varpi_v = \text{col}\{\varpi_{v1}, \varpi_{v2}, \ldots, \varpi_{vN}\}$.

Remark 5.3 A fault-tolerant hierarchical control method based on a virtual system is proposed for the complex decoupling problem between time-varying PLOE faults and adaptive controller gains under a directed communication network, as shown in Fig. 5.2. The upper layer is based on the virtual system $\hat{v}_{bdi}$ of neighborhood state information so that all virtual systems $\hat{v}_{bdi}$ track into the convex packet formed by the leaders v_{N+1}, v_{N+2}. The lower layer is designed with a decentralized adaptive tracking controller to achieve synchronization between the virtual system and the actual system.

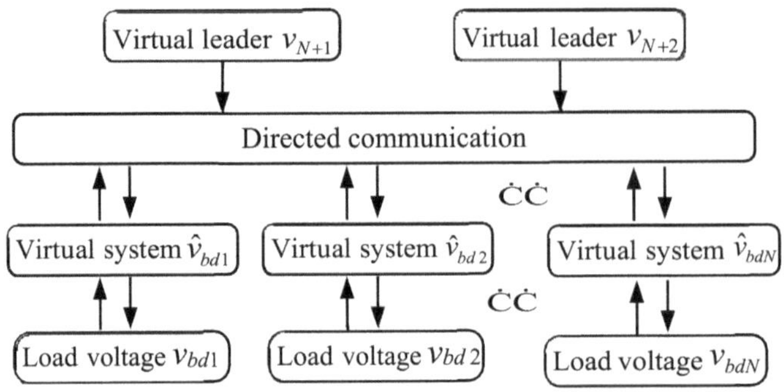

Fig. 5.2 Block diagram of a multi-MG system and the proposed containment-based secondary voltage regulation strategy

Theorem 5.1 *Under Assumption 5.3, the containment error of the load voltage value of DGs will converge exponentially to a bounded domain.*

Proof. Consider the following Lyapunov function candidate

$$V = V_1 + V_2, \tag{5.28}$$

where

$$V_1 = \sum_{i=1}^{N} \varpi_{vi}^2 + \sum_{i=1}^{N} \frac{\tilde{c}_{vi}^2}{\gamma_{c_{vi}}} + tr(\tilde{W}_v^T \Gamma_v^{-1} \tilde{W}_v), \tag{5.29}$$

and

$$V_2 = \frac{1}{2} \sum_{i=1}^{N} \pi_i (2\kappa_{vi} + \ell_{vi})\ell_{vi} + \frac{1}{2} \sum_{i=1}^{N} \pi_i \tilde{\kappa}_{vi}^2, \tag{5.30}$$

where $\tilde{W}_v = \text{diag}\{\tilde{W}_{v1}, \tilde{W}_{v2}, \ldots, \tilde{W}_{vN}\}$, $\tilde{W}_{vi} = \hat{W}_{vi} - W_{vi}$, $\Gamma_v = \text{diag}\{\gamma_{wv1}, \gamma_{wv2}, \ldots, \gamma_{wv3}\}$, $\tilde{c}_{vi} = c_{vi} - \bar{c}_{vi}$, and $\bar{c}_{vi}$ is a positive constant to be designed. Define $c_{vi}(0) \geq 0$, $\kappa_{vi}(0) \geq 1$, it is easy to obtain that $c_{vi}(t) \geq 0$, $\kappa_{vi}(t) \geq 1$ for any $t > 0$, respectively. It is not difficult to see that $V > 0$.

Take the derivative of V_1, from (5.23), we obtain that

$$\dot{V}_1 = 2 \sum_{i=1}^{N} \varpi_{vi} \dot{\varpi}_{vi} + \sum_{i=1}^{N} \frac{2\tilde{c}_{vi}\dot{c}_{vi}}{\gamma_{c_{vi}}} + 2tr(\tilde{W}^T \Gamma_W^{-1} \dot{\tilde{W}}). \tag{5.31}$$

Define $\tilde{\varphi} = \text{col}\{\tilde{\varphi}_1, \tilde{\varphi}_2, \ldots, \tilde{\varphi}_N\}$, $\tilde{\varphi}_{vi} = \varphi_{vi}(\hat{x}_{vi}) - \varphi_{vi}(x_{vi})$,

$$
\begin{aligned}
&2\sum_{i=1}^{N} \varpi_{vi}\dot{\varpi}_{vi} + \sum_{i=1}^{N} \frac{2\tilde{c}_{vi}\dot{c}_{vi}}{\gamma_{c_{vi}}} \\
&= -2\sum_{i=1}^{N}(1-\mathcal{Q}_{vi})\varpi_{vi}(\kappa_{vi}+\ell_{vi})\vartheta_{vi}(t) - 2\sum_{i=1}^{N} c_{vi}\mathcal{Q}_{vi}\varpi_{vi}\eta(\varpi_{vi}(t)) \\
&\quad - 2\sum_{i=1}^{N}\varpi_{vi}\mathcal{P}_{vi} - 2\sum_{i=1}^{N} l_{vi}\varpi_{vi}^2 + 2\sum_{i=1}^{N}\varpi_{vi}(\hat{W}_{vi}^T\varphi_{vi}(\hat{v}_{bdi}) \\
&\quad - \sum_{i=1}^{N} W_{vi}^T\varphi_{vi}(v_{bdi})) - 2\sum_{i=1}^{N}\varpi_{vi}o_{vi} + 2\sum_{i=1}^{N}(c_{vi}-\bar{c}_{vi})\|\varpi_{vi}\| \\
&\quad - 2\sum_{i=1}^{N}\frac{k_{c_{vi}}}{\gamma_{c_{vi}}}(c_{vi}-\bar{c}_{vi})c_{vi}.
\end{aligned}
\tag{5.32}
$$

Using Young's inequality, for any positive constants ψ_1, one has

$$
\begin{aligned}
&-2\sum_{i=1}^{N}(1-\mathcal{Q}_{vi})\varpi_{vi}(\kappa_{vi}+\ell_{vi})\vartheta_{vi} \\
&\quad \le \psi_1^{-1}\sum_{i=1}^{N}\varpi_{vi}^2 + \psi_1\sum_{i=1}^{N}(1-\mathcal{Q}_{vi})^2(\kappa_{vi}+\ell_{vi})^2\vartheta_{vi}^2.
\end{aligned}
\tag{5.33}
$$

Note that if $\|\varpi_{vi}(t)\| > \chi_{vi}, i = 1, 2, \ldots, N$.

$$
\varpi_{vi}(t)\eta(\varpi_{vi}(t)) = \|\varpi_{vi}(t)\| + c_{vi}\varpi_{vi}^2. \tag{5.34}
$$

Then, $\bar{c}_{vi} > 0$ to be sufficiently large such that $\bar{c}_{vi} \ge \varsigma_{vi}$, we can obtain

$$
\begin{aligned}
&-2\sum_{i=1}^{N}\varpi_{vi}c_{vi}\mathcal{Q}_{vi}\eta(\varpi_{vi}(t)) - 2\sum_{i=1}^{N}\varpi_{vi}\mathcal{P}_{vi} + 2\sum_{i=1}^{N}(c_{vi}-\bar{c}_{vi})\|\varpi_{vi}\| \\
&\le 2\sum_{i=1}^{N}(\varsigma_{vi}-c_{vi}\mathcal{Q}_{vi})\|\varpi_{vi}\| + 2\sum_{i=1}^{N}(c_{vi}-\bar{c}_{vi})\|\varpi_{vi}\| - 2\sum_{i=1}^{N}\mathcal{Q}_{vi}c_{vi}^2\|\varpi_{vi}\|^2 \\
&\le 2\sum_{i=1}^{N}(1-\mathcal{Q}_{vi})c_{vi}\|\varpi_{vi}\| - 2\sum_{i=1}^{N}\mathcal{Q}_{vi}c_{vi}^2\|\varpi_{vi}\|^2.
\end{aligned}
\tag{5.35}
$$

Note that if $\|\varpi_{vi}(t)\| < \chi_{vi}, i = 1, 2, \ldots, N$.

$$\begin{aligned}
&-2\sum_{i=1}^{N}\varpi_{vi}c_{vi}\mathcal{Q}_{vi}\eta(\varpi_{vi}(t)) - 2\sum_{i=1}^{N}\varpi_{vi}\mathcal{P}_{vi} + 2\sum_{i=1}^{N}(c_{vi}-\bar{c}_{vi})\|\varpi_{vi}\| \\
&\leq 2\sum_{i=1}^{N}(c_{vi}+\varsigma_{vi})\chi_{vi} - 2\sum_{i=1}^{N}\mathcal{Q}_{vi}c_{vi}^{2}\|\varpi_{vi}\|^{2}.
\end{aligned} \tag{5.36}$$

From (5.35) to (5.36), and define $\psi_2 \leq \frac{2\mathcal{Q}_{vi}}{(1-\mathcal{Q}_{vi})^2}$, we can obtain

$$\begin{aligned}
&-2\sum_{i=1}^{N}\varpi_{vi}c_{vi}\mathcal{Q}_{vi}\eta(\varpi_{vi}(t)) - 2\sum_{i=1}^{N}\varpi_{vi}\mathcal{P}_{vi} + 2\sum_{i=1}^{N}(c_{vi}-\bar{c}_{vi})\|\varpi_{vi}\| \\
&\leq 2\sum_{i=1}^{N}(1-\mathcal{Q}_{vi})c_{vi}\|\varpi_{vi}\| + 2\sum_{i=1}^{N}(c_{vi}+\varsigma_{vi})\chi_{vi} - 2\sum_{i=1}^{N}\mathcal{Q}_{vi}c_{vi}^{2}\|\varpi_{vi}\|^{2} \\
&\leq \sum_{i=1}^{N}\psi_2(1-\mathcal{Q}_{vi})^2c_{vi}^{2}\|\varpi_{vi}\|^{2} + \sum_{i=1}^{N}\psi_2^{-1} + 2\sum_{i=1}^{N}(c_{vi}+\varsigma_{vi})\chi_{vi} \\
&\quad -2\sum_{i=1}^{N}c_{vi}^{2}\mathcal{Q}_{vi}\|\varpi_{vi}\|^{2} \\
&\leq \psi_2^{-1}N + 2\sum_{i=1}^{N}(c_{vi}+\varsigma_{vi})\chi_{vi}.
\end{aligned} \tag{5.37}$$

Moreover, for any positive constants ψ_3, ψ_4, it is easy to obtain

$$-2\sum_{i=1}^{N}\varpi_{vi}o_{vi} \leq \psi_3^{-1}\sum_{i=1}^{N}\varpi_{vi}^{2} + \psi_3\sum_{i=1}^{N}o_{vi}^{2}, \tag{5.38}$$

and

$$\begin{aligned}
&2\sum_{i=1}^{N}\varpi_{vi}(\hat{W}_{vi}^{T}\varphi(\hat{v}_{bdi}) - W_{vi}^{T}\varphi(v_{bdi})) \\
&= 2\sum_{i=1}^{N}\varpi_{vi}W_{vi}^{T}\tilde{\varphi}_{v} + 2\sum_{i=1}^{N}\varpi_{vi}\tilde{W}_{vi}^{T}\varphi_{vi}(\hat{v}_{bdi}) \\
&\leq \psi_4^{-1}\sum_{i=1}^{N}\varpi_{vi}^{2} + \psi_4\sum_{i=1}^{N}\tilde{\varphi}_{vi}^{T}W_{vi}W_{vi}^{T}\tilde{\varphi}_{vi} + 2\sum_{i=1}^{N}\varpi_{vi}\tilde{W}_{vi}^{T}\varphi_{vi}(\hat{v}_{bdi}).
\end{aligned} \tag{5.39}$$

From (5.22), for any positive constant ψ_5, one can obtain that

$$\begin{aligned}
&2tr(\tilde{W}^T\Gamma_W^{-1}\dot{\tilde{W}}) \\
&\leq -2\sum_{i=1}^{N} tr(\tilde{W}_{vi}^T\varphi_{vi}(\hat{v}_{bdi})j_{vi}\varpi_{vi}) - 2\sum_{i=1}^{N} tr(\tilde{W}_{vi}^T\hat{W}_{vi}) \\
&\leq \sum_{i=1}^{N}\psi_5\breve{W}_{vi}^2 - \sum_{i=1}^{N}(2-\frac{1}{\psi_5})\|\tilde{W}_{vi}\|_F^2 - 2\sum_{i=1}^{N} tr(\tilde{W}_{vi}^T\varphi_{vi}(\hat{v}_{bdi})j_{vi}\varpi_{vi}). \quad (5.40)
\end{aligned}$$

Then, because of $j_{vi} \geq 1$,

$$\begin{aligned}
\dot{V}_1 \leq &(\psi_1^{-1}+\psi_3^{-1}+\psi_4^{-1}-2l_{vi})\sum_{i=1}^{N}\varpi_{vi}^2 + \psi_1\sum_{i=1}^{N}(1-\mathcal{Q}_{vi})^2(\kappa_{vi}+\ell_{vi})^2\vartheta_{vi}^2 \\
&+2\sum_{i=1}^{N}(c_{vi}+\varsigma_{vi})\chi_{vi} + \psi_2^{-1}N + \psi_3\sum_{i=1}^{N}o_{vi}^2 + \sum_{i=1}^{N}(4\psi_4\varphi_{mi}^2+\psi_5)\breve{W}_{vi}^2 \\
&-\sum_{i=1}^{N}(2-\frac{1}{\psi_5})\|\tilde{W}_{vi}\|_F^2 - 2\sum_{i=1}^{N}\frac{k_{c_{vi}}}{\gamma_{c_{vi}}}(c_{vi}-\bar{c}_{vi})c_{vi}. \quad (5.41)
\end{aligned}$$

From (5.27), we can obtain

$$\begin{aligned}
\dot{V}_2 &= \sum_{i=1}^{N}\pi_i(\kappa_{vi}+\ell_{vi})\dot{\ell}_{vi} + \sum_{i=1}^{N}\pi_i(\ell_{vi}+\kappa_{vi}-\bar{\kappa}_{vi})\dot{\kappa}_{vi} \\
&= -2\vartheta_v^T[(\kappa_v+\ell_v)\Pi\mathfrak{L}_1(\kappa_v+\ell_v)]\vartheta_v + 2\vartheta_v^T(\kappa_v+\ell_v)\Pi\mathfrak{L}_1\varphi^T(\hat{v}_{bd})\hat{W} \\
&\quad -2\vartheta^T(\kappa_v+\ell_v)\Omega\mathfrak{L}_1 L\varpi_v + \vartheta_v^T(\kappa_v+\ell_v-\bar{\kappa}_v)\Pi\vartheta_v \\
&\quad -\sum_{i=1}^{N}\pi_i\nu_{\kappa i}(\kappa_{vi}+\ell_{vi}-\bar{\kappa}_{vi})(\kappa_{vi}-1), \quad (5.42)
\end{aligned}$$

where $\bar{\kappa}_v = \mathrm{diag}\{\bar{\kappa}_{v1}, \bar{\kappa}_{v2}, \ldots, \bar{\kappa}_{vN}\}$. By Lemma 5.1, one obtains

$$\begin{aligned}
&-2\vartheta_v^T[(\kappa_v+\ell_v)\Pi\mathfrak{L}_1(\kappa_v+\ell_v)]\vartheta_v \\
&= -\vartheta_v^T[(\kappa_v+\ell_v)(\Pi\mathfrak{L}_1+\mathfrak{L}_1^T\Pi)(\kappa_v+\ell_v)]\vartheta_v \\
&\leq -\lambda_0\sum_{i=1}^{N}(\kappa_{vi}+\ell_{vi})^2\vartheta_{vi}^2. \quad (5.43)
\end{aligned}$$

For any positive constants ψ_6, ψ_7, it follows

$$
\begin{aligned}
&2\vartheta_v^T(\kappa_v+\ell_v)\varPi\mathfrak{L}_1\hat{W}_v^T\varphi(\hat{\upsilon}_{bd})\\
&\le \psi_6\lambda_{\max}(\varPi\mathfrak{L}_1\mathfrak{L}_1^T\varPi)\sum_{i=1}^{N}(\kappa_{vi}+\ell_{vi})^2\vartheta_{vi}^2+\frac{1}{\psi_6}\sum_{i=1}^{N}\varphi^T(\hat{\upsilon}_{bdi})\hat{W}_{vi}\hat{W}_{vi}^T\varphi(\hat{\upsilon}_{bdi}),
\end{aligned}
\tag{5.44}
$$

and

$$
\begin{aligned}
&-2\vartheta_v^T(\hat{\kappa}_v+\ell_v)\varPi\mathfrak{L}_1L\varpi_v\\
&\le \psi_7\lambda_{\max}(\varPi\mathfrak{L}_1\mathfrak{L}_1^T\varPi)\sum_{i=1}^{N}l_{vi}^2(\kappa_{vi}+\ell_{vi})^2\vartheta_{vi}^2+\frac{1}{\psi_7}\sum_{i=1}^{N}\varpi_{vi}^2.
\end{aligned}
\tag{5.45}
$$

Substituting (5.43)–(5.45) into (5.42), and from (5.41), we obtain

$$
\begin{aligned}
\dot{V}\le&(\psi_1^{-1}+\psi_3^{-1}+\psi_4^{-1}+\psi_7^{-1}-2l_{vi})\sum_{i=1}^{N}\varpi_{vi}^2\\
&-\sum_{i=1}^{N}[\lambda_0-\psi_1\sum_{i=1}^{N}(1-\mathcal{Q}_{vi})^2-(\psi_7l_{vi}^2+\psi_6)\lambda_{max}(\varPi\mathfrak{L}_1\mathfrak{L}_1^T\varPi)]\\
&\times(\kappa_{vi}+\ell_{vi})^2\vartheta_{vi}^2+\vartheta_v^T(\kappa_v+\ell_v-\bar{\kappa}_v)\varPi\vartheta_v+\psi_2^{-1}N+\psi_8\sum_{i=1}^{N}c_{vi}^2\\
&+2\sum_{i=1}^{N}(c_{vi}+\varsigma_{vi})\chi_{vi}+\psi_3\sum_{i=1}^{N}o_{vi}^2+\sum_{i=1}^{N}[(4\psi_4+\frac{1}{\psi_6})\varphi_{mi}^2+\psi_5]\breve{W}_{vi}^2\\
&-\sum_{i=1}^{N}\left(2-\frac{1}{\psi_5}\right)\|\tilde{W}_{vi}\|_F^2-2\sum_{i=1}^{N}\frac{k_{c_{vi}}}{\gamma_{c_{vi}}}(c_{vi}-\bar{c}_{vi})\,c_{vi}\\
&-\sum_{i=1}^{N}\pi_i\nu_{\kappa i}(\kappa_{vi}+\ell_{vi}-\bar{\kappa}_{vi})(\kappa_{vi}-1).
\end{aligned}
\tag{5.46}
$$

Note that

$$
\begin{aligned}
-(\kappa_{vi}-1)(\kappa_{vi}-\bar{\kappa}_{vi})&=-(\kappa_{vi}-\bar{\kappa}_{vi})^2-(\bar{\kappa}_{vi}-1)(\kappa_{vi}-\bar{\kappa}_{vi})\\
&\le-\frac{1}{2}(\kappa_{vi}-\bar{\kappa}_{vi})^2+\frac{1}{2}(\bar{\kappa}_{vi}-1)^2,
\end{aligned}
\tag{5.47}
$$

and

$$
\begin{aligned}
&-(\kappa_{vi}-1)(\kappa_{vi}-\bar{\kappa}_{vi})=-(\kappa_{vi}-1)^2-(\bar{\kappa}_{vi}-1)(\kappa_{vi}-\bar{\kappa}_{vi})\\
&\le-\frac{1}{2}(\kappa_{vi}-1)^2+\frac{1}{2}(\bar{\kappa}_{vi}-1)^2.
\end{aligned}
\tag{5.48}
$$

Furthermore

$$
\begin{aligned}
&-\sum_{i=1}^{N}\pi_i \nu_{\kappa_{vi}}(\kappa_{vi}-\bar{\kappa}_{vi})(\kappa_{vi}-1)\\
&=\sum_{i=1}^{N}\frac{\pi_i \nu_{\kappa_{vi}}}{2}(\bar{\kappa}_{vi}-1)^2-\frac{\pi_i \nu_{\kappa_{vi}}}{4}\left[(\kappa_{vi}-\bar{\kappa}_{vi})^2+(\kappa_{vi}-1)^2\right].
\end{aligned} \tag{5.49}
$$

Similarly,

$$
\begin{aligned}
&-\sum_{i=1}^{N}(c_{vi}-\bar{c}_{vi})k_{c_{vi}}c_{vi}\\
&=\sum_{i=1}^{N}\frac{k_{c_{vi}}}{2}\bar{c}_{vi}^2-\sum_{i=1}^{N}\frac{k_{c_{vi}}}{4}\left[(c_{vi}-\bar{c}_{vi})^2+c_{vi}^2\right].
\end{aligned} \tag{5.50}
$$

Moreover,

$$
2\sum_{i=1}^{N}(c_{vi}+\varsigma_{vi})\chi_{vi}\le\psi_8\sum_{i=1}^{N}c_{vi}^2+\sum_{i=1}^{N}(\psi_8^{-1}\chi_{vi}^2+2\varsigma_{vi}\chi_{vi}). \tag{5.51}
$$

Since, ψ_1,ψ_6 and ψ_7 to satisfy $\jmath=\lambda_0-\psi_1\sum_{i=1}^{N}(1-\mathcal{Q}_{vi})^2-(\psi_7 l_{vi}^2+\psi_6)\lambda_{\max}(\Pi\mathfrak{L}_1\mathfrak{L}_1^T\Pi)\ge 0$, and $\bar{\kappa}_{vi}\ge\varpi_{vi}/\jmath$, then

$$
\begin{aligned}
\dot{V}\le&-\delta V-\left(\frac{1}{2}-\delta\right)\vartheta_v^T(\kappa_v+\ell_v)\Pi\vartheta_v\\
&+(\psi_1^{-1}+\psi_3^{-1}+\psi_4^{-1}+\psi_7^{-1}+\delta-2l_{vi})\sum_{i=1}^{N}\varpi_{vi}^2+\psi_2^{-1}N+\psi_3\sum_{i=1}^{N}o_{vi}^2\\
&+\sum_{i=1}^{N}[(4\psi_4+\frac{1}{\psi_6})\varphi_{mi}^2+\psi_5]\breve{W}_{vi}^2-\sum_{i=1}^{N}(2-\frac{1}{\psi_5}-\gamma_{wvi}\delta)\|\tilde{W}_{vi}\|_F^2\\
&+\sum_{i=1}^{N}\pi_i\nu_{\kappa_{vi}}(\bar{\kappa}_{vi}-1)^2-\sum_{i=1}^{N}\left(\frac{\nu_{\kappa_{vi}}}{2}-\frac{1}{2}\delta\right)\pi_i\tilde{\kappa}_{vi}^2+\sum_{i=1}^{N}k_{c_{vi}}\bar{c}_{vi}^2\\
&-\left(\frac{k_{c_{vi}}}{2}-\frac{\delta}{\gamma_{c_{vi}}}\right)\tilde{c}_{vi}^2-\left(\frac{k_{c_{vi}}}{2}-\psi_8\right)c_{vi}^2+\sum_{i=1}^{N}\left(\psi_8^{-1}\chi_{vi}^2+2\varsigma_{vi}\chi_{vi}\right).
\end{aligned} \tag{5.52}
$$

Moreover, when $0<\delta\le\min\{h,\frac{1}{2},\frac{2}{\gamma_{wvi}}-\frac{1}{\psi_5\gamma_{wvi}},\nu_{\kappa_{vi}},\frac{k_{c_{vi}}\gamma_{c_{vi}}}{2}\}$ and $\psi_8\le\frac{k_{c_{vi}}}{2}$, $h=2l_{vi}-\psi_1^{-1}-\psi_3^{-1}-\psi_4^{-1}-\psi_7^{-1}>0$ such that (5.38) satisfies the following inequality:

$$\dot{V} \leq -\delta V + \Xi_v, \tag{5.53}$$

where

$$\Xi_v = \psi_2^{-1} N + \psi_3 \sum_{i=1}^{N} o_{vi}^2 + \sum_{i=1}^{N} \left[\left(4\psi_4 + \frac{1}{\psi_6} \right) \varphi_{mi}^2 + \psi_5 \right] \breve{W}_{vi}^2 \\ + \sum_{i=1}^{N} \frac{\pi_i \nu_{\kappa_{vi}}}{2} (\bar{\kappa}_{vi} - 1)^2 + \sum_{i=1}^{N} \frac{k_{c_{vi}}}{2} \bar{c}_{vi}^2 + \sum_{i=1}^{N} \left(\psi_8^{-1} \chi_{vi}^2 + 2\varsigma_{vi} \chi_{vi} \right). \tag{5.54}$$

V exponentially converges to the bouded set

$$\mathcal{D} \triangleq \left\{ \varpi_v, \tilde{c}_v, \tilde{W}_v, \vartheta_v, \tilde{\kappa}_v : V \leq \frac{1}{\delta} \Xi_v \right\}. \tag{5.55}$$

From (5.55), we can obtain that ϑ_v can converge exponentially to the residual set $\mathcal{D}$ and converge faster than $e^{-\delta t}$.

■

Remark 5.4 In this chapter, all information related to the fault parameters is unknown, including quantitative information on boundaries, frequencies and types (whether they are constant, time-varying, periodic, etc.). It is well known that the fault parameters can be different for different agents.

Remark 5.5 Due to the introduction of adaptive dynamics $\rho_{vi}(t), \vartheta_{vi}, \iota_{vi}$ in the control protocol, the protocol no longer needs to include information related to parameters such as the communication network structure and the number of DGs. The protocol is therefore implemented in a fully distributed framework. This helps to reduce the bandwidth requirements and costs of the underlying communication network and enhance the flexibility and reliability of the overall MG operation.

Remark 5.6 The adaptive and bounded layers introduced in the strategy help overcome the limitations of fault factors and the undesirable effects of non-smooth controllers in [11].

5.3.2 The Consensus-Based Frequency Restoration

The fault model for input $u_{\omega i}$ of the ith DG can be represented as follows:

$$\dot{\omega}_i = \mathcal{Q}_{\omega i} u_{\omega i} + \mathcal{P}_{\omega i}, i = 1, \ldots, N. \tag{5.56}$$

Let the controller $u_{\omega i}$ in (5.56) be

$$u_{\omega i} = -(\kappa_{\omega i} + \ell_{\omega i}) \vartheta_{\omega i}(t) + c_{\omega i} \eta(\varpi_{\omega i}(t)), \tag{5.57}$$

where

$$\dot{\kappa}_{\omega i} = -\nu_{\kappa_{\omega i}}(\kappa_{\omega i} - 1) + \vartheta_{\omega i}^2(t), \tag{5.58}$$

and

$$\ell_{\omega i} = \vartheta_{\omega i}^2(t), \tag{5.59}$$

where

$$\dot{c}_{\omega i} = -k_{c_{\omega i}} c_{\omega i} + \gamma_{c_{\omega i}} \|\varpi_{\omega i}\|, \tag{5.60}$$

$$\eta(*) = \begin{cases} \frac{*}{\|*\|} + c_{\omega i}*, & if \ \| * \| > \chi_{\omega i} \\ \frac{*}{\chi_{\omega i}} + c_{\omega i}*, & if \ \| * \| \leq \chi_{\omega i}. \end{cases} \tag{5.61}$$

Then, we have

$$\dot{\omega}_i = -\mathcal{Q}_{\omega i}(\kappa_{\omega i} + \ell_{\omega i})\vartheta_{\omega i}(t) + c_{\omega i}\mathcal{Q}_{\omega i}\eta(\varpi_{\omega i}(t)) + \mathcal{P}_{\omega i}, i = 1, \ldots, N. \tag{5.62}$$

Define the following parameters:

$$\vartheta_{\omega i} = \sum_{j=1}^{N} a_{ij}(\hat{\omega}_i - \hat{\omega}_j) + g_{\omega i}(\hat{\omega}_i - \omega_{ref}), i = 1, \ldots, N. \tag{5.63}$$

The virtual system $\hat{\omega}_i$ to estimate $v_{bi}, i = 1, 2, \ldots, N$ as follows:

$$\dot{\hat{\omega}}_i = \hat{u}_{\omega i} - l_{vi}(\hat{\omega}_i - \omega_i). \tag{5.64}$$

The controller of the virtual system can be designed as

$$\hat{u}_{\omega i} = -(\kappa_{\omega i} + \ell_{\omega i})\vartheta_{\omega i}(t). \tag{5.65}$$

5.3.3 *The Consensus-Based Active Power Control*

The fault model for input u_{Pi} of the ith DG can be represented as follows:

$$m_i \dot{P}_i = \mathcal{Q}_{Pi} u_{Pi} + \mathcal{P}_{Pi}, i = 1, \ldots, N. \tag{5.66}$$

Let the controller u_{Pi} in (5.66) be

$$u_{Pi} = -(\kappa_{Pi} + \ell_{Pi})\vartheta_{Pi}(t) + c_{Pi}\eta(\varpi_{Pi}(t)), \tag{5.67}$$

where

$$\dot{\kappa}_{Pi} = -\nu_{\kappa_{Pi}}(\kappa_{Pi} - 1) + \vartheta_{Pi}^2(t), \tag{5.68}$$

and

$$\ell_{Pi} = \vartheta_{Pi}^2(t), \tag{5.69}$$

where

$$\dot{c}_{Pi} = -k_{c_{Pi}} c_{Pi} + \gamma_{c_{Pi}} \|\varpi_{Pi}\|, \tag{5.70}$$

$$\eta(*) = \begin{cases} \frac{*}{\|*\|} + c_{Pi}*, & if\ \| * \| > \chi_{Pi} \\ \frac{*}{\chi_{Pi}} + c_{Pi}\breve{W}_{vi}^2, & if\ \| * \| \le \chi_{Pi}. \end{cases} \tag{5.71}$$

Then, we have

$$m_i\dot{P}_i = -\mathcal{Q}_{Pi}(\kappa_{Pi} + \ell_{Pi})\vartheta_{Pi}(t) + c_{Pi}\mathcal{Q}_{Pi}\eta(\varpi_{Pi}(t)) + \mathcal{P}_{Pi}, i = 1, \ldots, N. \tag{5.72}$$

Define the following parameters:

$$\vartheta_{Pi} = \sum_{j=1}^{N} a_{ij}(m_i\hat{P}_i - m_j\hat{P}_j), i = 1, \ldots, N. \tag{5.73}$$

The virtual system $\hat{\upsilon}_{bi}$ to estimate υ_{bi}, $i = 1, 2, \ldots, N$ as follwes:

$$m_i\dot{\hat{P}}_i = \hat{u}_{Pi} - l_{vi}(m_i\hat{P}_i - m_iP_i). \tag{5.74}$$

The controller of the virtual system can be designed as

$$\hat{u}_{Pi} = -(\kappa_{Pi} + \ell_{Pi})\vartheta_{Pi}(t). \tag{5.75}$$

5.4 Simulation Results

In order to demonstrate the feasibility and effectiveness of the proposed containment-based load voltage regulate strategy and the consensus-based frequency and active power-based control strategy, we conducted simulations of various MG scenarios using the MATLAB/Simulink environment, as depicted in Fig. 5.3. The MG system was tested with a combination of four DGs, with v_{bd1}, v_{bd2}, v_{bd3}, and v_{bd4} representing the bus voltage of DG#1, DG#2, DG#3, and DG#4, respectively. Table 5.1 presents the parameters of the DGs, lines, and loads, which can be found in Table 5.1. It is assumed that the DGs communicate with each other via the directed topology shown in Fig. 5.3, with DG#1 and DG#4 receiving information from v_5 and v_6, respectively.

We define $v_5(0) = 300$, $v_6(0) = 320$. By calculating that we get $\bar{v}_{bd} = \text{col}\{310, 312.5, 315, 317.5\}$. Let $\kappa_{vi}(0.7) = 100$ and $c_{vi}(0.7) = 18$ be the initial value of $\kappa_{vi}(t)$ and $c_{vi}(t)$, $i = 1, 2, \ldots, 4$, respectively. The actuator faults are chosen as: $\mathcal{Q}_{v1} =$

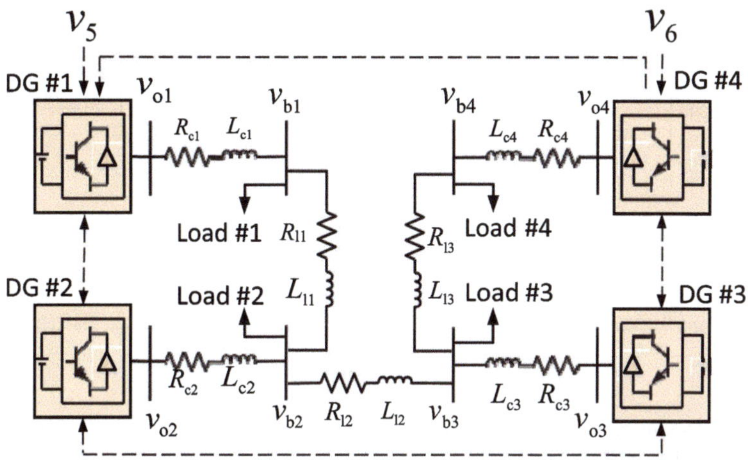

Fig. 5.3 Block diagram of a multi-MG system and the proposed containment-based secondary voltage regulation strategy

Table 5.1 Specifications of the MG test system

DGs	DG#1 & DG#2 & DG#3 & DG#4		
m_P	1.5×10^{-5}		
n_Q	2×10^{-4}		
Lines	$R_{l1} = R_{l3} = 1e-4\,\Omega$	$R_{l2} = 1e-4\,\Omega$	
	$L_{l1} = L_{l3} = 0.318\,\text{emH}$	$L_{l2} = 1.847\,\text{mH}$	
RL loads	load #1 & load #2	load #3	& load #4
P(per phase) kW	50	5	0.5
Q(per phase) kVAr	12	1.2	0.1

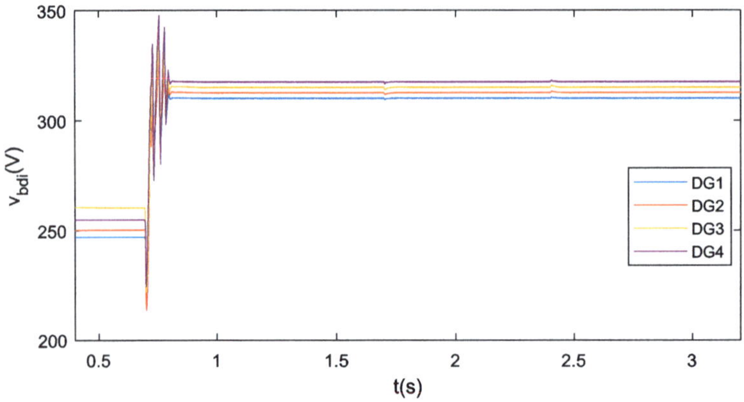

Fig. 5.4 Block diagram of a multi-MG system and the proposed containment-based secondary voltage regulation strategy

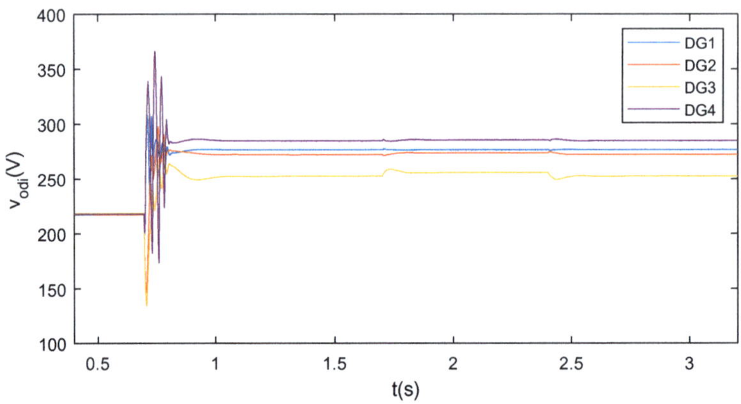

Fig. 5.5 Block diagram of a multi-MG system and the proposed containment-based secondary voltage regulation strategy

$\mathcal{Q}_{\omega 1} = \mathcal{Q}_{P1} = 0.5\cos(t+10)+0.1,\ \mathcal{Q}_{v2} = \mathcal{Q}_{\omega 2} = \mathcal{Q}_{P2} = \begin{cases} 0.5\sin(2t-1)+0.3, t \leq 8 \\ 0.8, t > 8 \end{cases},$
$\mathcal{Q}_{v3} = \mathcal{Q}_{\omega 3} = \mathcal{Q}_{P3} = \begin{cases} 0.9, t \leq 5 \\ 0.8\cos(6t+3)+0.1, t > 5 \end{cases}, \mathcal{Q}_{v4} = \mathcal{Q}_{\omega 4} = \mathcal{Q}_{P4} = 0.9, \mathcal{P}_{v1} =$
$\mathcal{P}_{\omega 1} = \mathcal{P}_{P1} = 0.1\cos(t-10)+0.2, \mathcal{P}_{v2} = \mathcal{P}_{\omega 2} = \mathcal{P}_{P2} = \begin{cases} 0.2\sin(t+2)+0.05, t \leq 8 \\ 0.3, t > 8 \end{cases},$
$\mathcal{P}_{v3} = \mathcal{P}_{\omega 3} = \mathcal{P}_{P3} = \begin{cases} 0.4, t \leq 5 \\ 0.1\sin(t+3)+0.2, t > 5 \end{cases}, \mathcal{P}_{v4} = \mathcal{P}_{\omega 4} = \mathcal{P}_{P4} = 0.3.$

To verify the effectiveness of the proposed strategy, as illustrated in Figs. 5.4, 5.5, 5.6, 5.7 and 5.8, we initially relied solely on primary control. At t = 0.7 s, the sec-

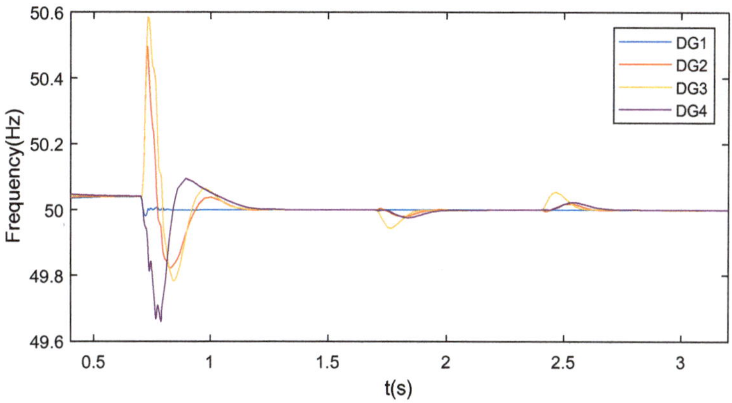

Fig. 5.6 Block diagram of a multi-MG system and the proposed containment-based secondary voltage regulation strategy

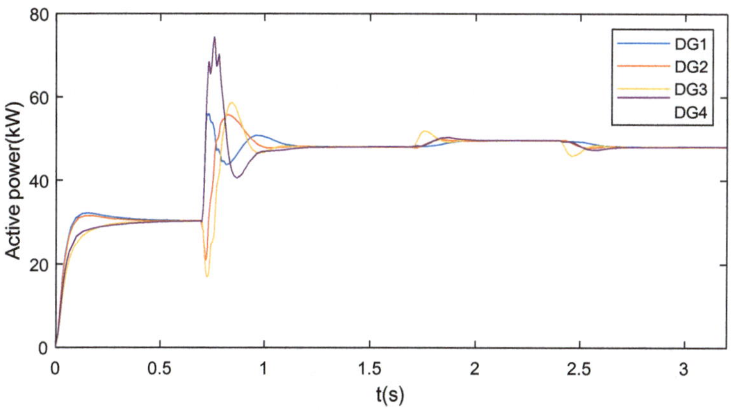

Fig. 5.7 Block diagram of a multi-MG system and the proposed containment-based secondary voltage regulation strategy

ondary control of the containment-based load voltage regulation and the consensus-based frequency and active power-based control came into effect. The load voltage is restored to a convex packet tensioned by all reference values, the frequency is restored to the reference value, and the active power is power shared. At t = 1.7 s, the additional load $P = 5\,\text{kW}$, $Q = 1.2\,\text{kVAr}$ (100% load increment) is attached to Load #3. Although there is some fluctuation in the load voltage, it quickly recovers to the convex packet tensed by all reference values. In the following scenario, the load increment is disconnected from the MG at 1.2 s. While there was some fluctuation in the load voltage, it quickly returned to the convex hull spanned by all reference values. The results of this simulation are depicted in Figs. 5.4, 5.5, 5.6, 5.7 and 5.8. The proposed control method effectively regulated load voltage, frequency, active power,

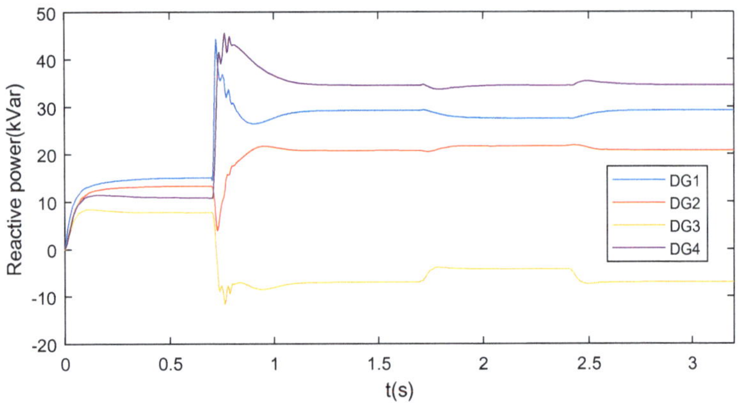

Fig. 5.8 Block diagram of a multi-MG system and the proposed containment-based secondary voltage regulation strategy

and reactive power in response to large-signal disturbances (i.e., load changes). It can be seen that the containment-based load voltage and consensus-based frequency fault-tolerant secondary control strategies proposed in this chapter are effective and can also be applied to load-switching scenarios.

5.5 Conclusions

This chapter proposed a fault-tolerant secondary control strategy for load voltage and frequency. Firstly, we transform the regulation problem of the load voltage into a containment control problem of a linear MAS containing unknown nonlinear dynamics, considering that the power flow between load buses depends on voltage difference. Next, we design a fault-tolerant secondary control strategy with hierarchical control. At the network layer, we create a virtual auxiliary system with adjacency information without faults and use a neural network to approximate the nonlinear dynamics. We use a fault compensator as a virtual actuator to adaptively reconfigure the faulty device in the physical layer. This layered process decouples asymmetric network and time-varying PLOE fault parameters, enabling fault-tolerant control of actuator PLOE faults and deviation faults in a fully distributed framework. Additionally, this strategy provides a feasible solution for industrial applications where the boundaries of the fault parameters can be unknown.

References

1. Bidram, A., Davoudi, A., Lewis, F.L., Guerrero, J.M.: Distributed cooperative secondary control of microgrids using feedback linearization. IEEE Trans. Power Syst. **28**(3), 3462–3470 (2013)
2. Dehkordi, N.M., Sadati, N., Hamzeh, M.: Fully distributed cooperative secondary frequency and voltage control of islanded microgrids. IEEE Trans. Energy Convers. **32**(2), 675–685 (2017)
3. Shahab, M.A., Mozafari, B., Soleymani, S., Dehkordi, N.M., Shourkaei, H.M., Guerrero, J.M.: Distributed consensus-based fault tolerant control of islanded microgrids. IEEE Trans. Smart Grid **11**(1), 37–47 (2020)
4. Afshari, A., Karrari, M., Baghaee, H.R., Gharehpetian, G.B.: Distributed fault-tolerant voltage/frequency synchronization in autonomous AC microgrids. IEEE Trans. Power Syst. **35**(5), 3774–3789 (2020)
5. Zhai, M., Sun, Q., Wang, R., Wang, B., Liu, S., Zhang, H.: Fully distributed fault-tolerant event-triggered control of microgrids under directed graphs. IEEE Trans. Netw. Sci. Eng. **9**(5), 3570–3579 (2022)
6. Zhai, M. ,Sun, Q. , Wang, B., Liu, Z., Zhang, H.: Cooperative fault-estimation-based event-triggered fault-tolerant voltage restoration in islanded AC microgrids. IEEE Trans. Autom. Sci. Eng. (2022)
7. Ning, B., Han, Q.L., Ding, L.: Distributed finite-time secondary frequency and voltage control for islanded microgrids with communication delays and switching topologies. IEEE Trans. Cybern. **51**(8), 3988–3999 (2021)
8. Afshari, A., Karrari, M., Baghaee, H.R., Gharehpetian, G.B., Karrari, S.: Cooperative fault-tolerant control of microgrids under switching communication topology. IEEE Trans. Smart Grid **11**(3), 1866–1879 (2020)
9. Dehkordi, N.M., Baghaee, H.R., Sadati, N., Guerrero, J.M.: Distributed noise-resilient secondary voltage and frequency control for islanded microgrids. IEEE Trans. Smart Grid **10**(4), 3780–3790 (2019)
10. Peng, Z., Wang, D., Zhang, H., Sun, G.: Distributed neural network control for adaptive synchronization of uncertain dynamical multiagent systems. IEEE Trans. Neural Netw. **25**(8), 1508–1519 (2014)
11. Luo, S., Ye, D.: Adaptive double event-triggered control for linear multi-agent systems with actuator faults. IEEE Trans. Circ. Syst. I: Regular Papers **66**(12), 4829–4839 (2019)

Chapter 6
Distributed Critical Bus Voltage Regulation Control for Multi-microgrids with Positive Minimum Inter-Event Times

Abstract This article proposes a distributed containment-based critical bus voltage coregulation strategy of submicrogrids (SMGs) with positive minimum interevent times (MIETs) in a multimicrogrid system (MMGS). First, the feedback linearization technique allows us to deal with nonlinear distributed generator (DG) dynamics. The critical bus voltage regulation problem of an ac microgrid (MG) is transformed into an output feedback tracking problem of a linear multiagent system (MAS) containing nonlinear dynamics. In addition, an event-triggered strategy based on containment control is proposed. The strategy limits each critical bus voltage within a reasonable range while having a voltage difference between SMGs to allow power flow. Furthermore, an innovative trigger mechanism has been designed based on multiple measurement errors. Based on this mechanism, the controllers are updated in an acyclic manner at the moment of event sampling, saving computational resources and transmission load. In addition, the proposed event-triggering mechanism guarantees the absence of Zeno behavior and a strictly positive MIET between each DG event. It is also more practical than the traditional event triggering because the hardware constraint always imposes a positive minimum time constraint between two consecutive event times. The protocol proposed in this article is fully distributed and scalable and does not depend on global information about the network graph. Finally, simulation results verify the effectiveness of the strategy.

Keywords AC microgrid · Adaptive control · Containment control · Event-triggered control · Voltage regulation

6.1 Introduction

In this chapter, we study the AC MG containment-based event-triggered critical bus voltage regulation strategy with MIET. The goal is the co-regulation problem of the critical bus voltage. The critical bus refers to the bus that connects the essential loads. The feedback linearization technique allows us to deal with nonlinear DG dynamics in a reduced-order approach. The unavailability of full-state measurements is considered. Thus, the critical bus voltage regulation problem of the AC MG is transformed

J. Sun et al., *Building a Low-Carbon Future*, Studies in Systems, Decision and Control 506, https://doi.org/10.1007/978-3-031-43498-3_6

into a distributed output feedback tracking problem of a linear MAS with nonlinear dynamics. This transformation is an indispensable preprocessing in the design of the control algorithm. Adjusting the critical bus voltages to reasonable values is necessary. However, when the bus voltages of all sub-MGs of a multi-MG system are controlled at their nominal values, the power flow will not be exchanged between the sub-MGs because of no voltage difference. For this reason, a containment-based distributed controller is proposed to balance the two conflicting objectives of voltage regulation and tide. This controller enables the critical bus voltage to be controlled within a reasonable range while a voltage difference exists to achieve power flow. Unlike the current event-based secondary control strategy, a suitable timer is designed to ensure the presence of a positive MIET for each DG unit. Moreover, this secondary controller relies only on information related to the observer, preventing the use of global information about the communication topology. The proposed algorithm innovatively relies on multiple measurement errors $\breve{e}_i$ and avoids continuous controller updates.

6.2 Modeling Framework

The considered a multi-MG system is composed of N series-connected sub-MGs. A multi-MG system is considered a MAS with sub-MGs as its agents. The sub-MGs can communicate with each other through the communication network $\mathbb{G}(\mathcal{C}, \mathcal{T})$. $\mathcal{C} = \{c_1, c_2, \ldots, c_N\}$ is the node set, and $\mathcal{T} \subseteq \mathcal{C} \times \mathcal{C}$ is the edge set. The weighted adjacency matrix $\mathcal{A} = [a_{ij}] \in \mathbb{R}^{(N+1)\times(N+1)}$ is defined by $a_{ij} = 1$, if $(c_i, c_j) \in \mathcal{T}$, otherwise, $a_{ij} = 0$. The Laplacian matrix $\mathfrak{L}$ satisfies $\mathfrak{L}_{ii} = \sum_{j=0}^{N} a_{ij}$ and $\mathfrak{L}_{ij} = -a_{ij}$, when $i \neq j$.

In the sub-MG, each DG unit, as shown in Fig. 5.1, contains inner-loop voltage and current power, LC filter, and line model. The nonlinear dynamic model of the AC MG in a compact form as

$$\begin{aligned} \dot{x}_i &= f_i(x_i) + k_i(x_i)D_i + g_i(x_i)u_i \\ \mathcal{Y}_i &= h_i(x_i), \end{aligned} \tag{6.1}$$

where the state vector is

$$x_i = [\delta_i, P_i, Q_i, \phi_{v_{di}}, \phi_{v_{qi}}, \phi_{i_{di}}, \phi_{i_{qi}}, i_{ldi}, i_{lqi}, v_{odi}, v_{oqi}, i_{odi}, i_{oqi}]^T, \tag{6.2}$$

and $D_i = [\omega_{com}, v_{bdi}, v_{bqi}]$. Detailed expressions for $f_i(x_i)$, $g_i(x_i)$, $k_i(x_i)$, and $h_i(x_i)$ can be extracted from (6.1)–(6.4). $\mathcal{Y}_i$ is set to v_{bi} which is represents the critical bus voltage.

By feedback linearization, there hold

$$\ddot{\mathcal{Y}}_i = L_{F_i}^2 h_i + L_{g_i} L_{F_i}(x_i) h_i V_{ni} + \ddot{v}_{li} = u_i + \ddot{v}_{li}, \tag{6.3}$$

where $F_i(x_i) = f_i(x_i) + k_i(x_i)D_i$, and $L^2_{F_i}h_i = L_{F_i}(L_{F_i}h_i) = \frac{\partial L_{F_i}h_i}{\partial x_i}F_i$. v_{li} denotes the line voltage between the critical bus and the inverter output.

It is difficult to obtain the value of $\dot{v}_{bdi}$ in practice, the design of the control algorithm using only v_{bdi} instead of r_i, where r_i is translated from (6.3) as follows:

$$\begin{cases} \dot{r}_i(t) = Ar_i(t) + B(u_i + \ddot{v}_{li}), i = 1, \ldots, N, \\ y_i(t) = Cr_i(t), \end{cases} \tag{6.4}$$

where $r_i = [v_{bdi}, \dot{v}_{bdi}]^T$, $A = \begin{bmatrix} 0 & 1 \\ 0 & 0 \end{bmatrix}$, $B = [0, 1]^T$, $C = [1, 0]$.

Remark 6.1 The feedback linearization technique allows us to treat nonlinear DG dynamics with a reduced-order approach. The difference is that we aim at the critical bus voltage regulation problem. The critical bus voltage regulation problem of an AC MG is transformed into a distributed output feedback tracking problem with the nonlinear dynamics of a linear MAS.

In order to regulate the voltage to a reasonable range and to enable the power flow between sub-MGs by making a voltage difference between them, we design a secondary voltage strategy for a multi-MG system based on containment control as shown in Fig. 6.1.

The following definition applies to the inclusion-based voltage controller, which specifies the lower and upper limits of the voltage regulation objectives. The boundary dynamics are as follows:

$$\dot{r}_m(t) = Ar_m(t), m = N + 1, N + 2, \tag{6.5}$$

where $r_m = [v_{ref_{m-N}}, 0]^T$. Define $r_{ref} = \text{col}\{r_{N+1}, x_{N+2}\}$, and v_{ref_1}, v_{ref_2} represent the upper and lower boundaries of the voltage regulation, respectively.

Assumption 6.1 The subgraph associated with the DGs is undirected. For each DG, at least one of the virtual leader's dynamics has paths to all DGs.

Remark 6.2 Assumption 6.1 is the basic standard assumption of MASs control, on which the following results are obtained.

Then, $\mathfrak{L}$ is defined as

$$\mathfrak{L} = \begin{bmatrix} 0 & 0_{2\times N} \\ \mathfrak{L}_2 & \mathfrak{L}_1 \end{bmatrix}, \tag{6.6}$$

where $\mathfrak{L}_2 \in \mathbb{R}^{N\times 2}$, $\mathfrak{L}_1 \in \mathbb{R}^{N\times N}$. The real part of each eigenvalue of $\mathfrak{L}_1$ is positive. It is easy to verify that $\mathcal{L}_1$ is a nonsingular M-matrix.

Under Assumption 6.1, $-\mathfrak{L}_1^{-1}\mathfrak{L}_2 r_d$ is within the convex hull spanned by the multiple leaders. Let

$$\bar{r} = -\mathfrak{L}_1^{-1}\mathfrak{L}_2 r_{ref}. \tag{6.7}$$

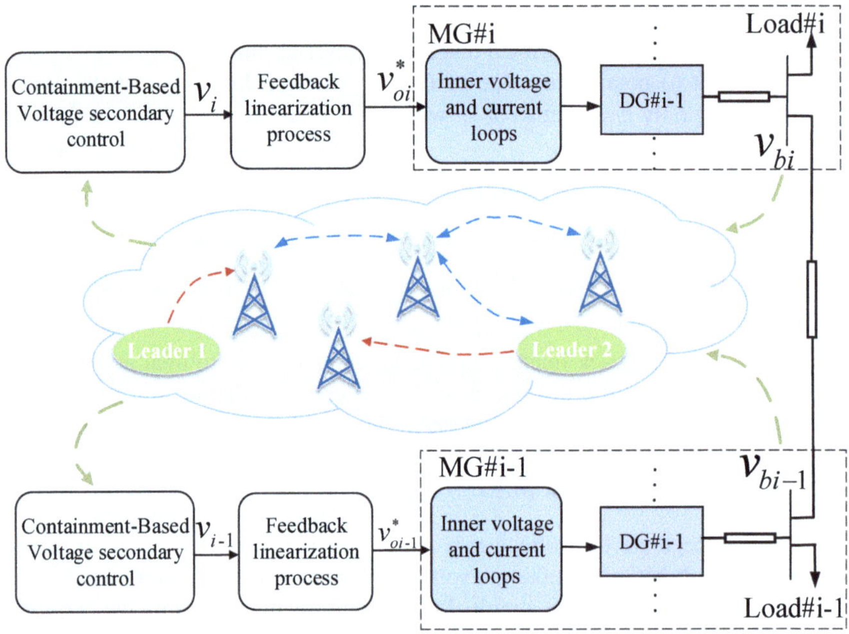

Fig. 6.1 Block diagram of a multi-MG system and the proposed containment-based secondary voltage regulation strategy

6.3 Main Result

6.3.1 Fuzzy Logic System

Then, we obtain

$$\ddot{v}_{li} = \varphi_i^T(r_i)\theta_i + o_i(t), \tag{6.8}$$

where $\theta_i = \text{col}\{\theta_{i1}, \theta_{i2}, \ldots, \theta_{im}\}$ with $\theta_{ik} \in \mathbb{R}^l$, $k = 1, 2, \ldots, m$ is the unknown parameter, $\varphi_i^T(r_i) = \text{diag}\{\varphi_{i1}^T, \varphi_{i2}^T, \ldots, \varphi_{im}^T\}$ with $\varphi_{ik} \in \mathbb{R}^l$ is the fuzzy basis function. $o_i \in \mathbb{R}^m$ with $\|o_i(t)\| \leq \bar{o}_i$ is a time-varying function, $\bar{o}_i$ is an unknown constant.

The dynamic model of the ith DG is

$$\begin{cases} \dot{r}_i(t) = Ar_i(t) + B(u_i + \varphi_i^T(r_i)\theta_i + o_i), \\ y_i(t) = Cr_i(t). \end{cases} \tag{6.9}$$

6.3.2 The Containment-Based Critical Bus Voltage Regulation Control with MIET

The state observer $\hat{r}_i$ to estimate r_i, $i = 1, 2, \ldots, N$ as follwes:

$$\begin{cases} \dot{\hat{r}}_i(t) = A\hat{r}_i(t) + B(u_i + \varphi_i^T(\hat{r}_i)\hat{\theta}_i) - L(\hat{y}_i - y_i), \\ \hat{y}_i(t) = C\hat{r}_i(t), \end{cases} \tag{6.10}$$

where $\hat{r}_i(t)$ is the estimation of $r_i(t)$, $\hat{y}_i(t)$ is the observer output vector, $L \in \mathbb{R}^{2\times 1}$ is the observer gain. Moreover

$$\dot{\hat{\theta}}_i = \gamma_{\theta i}[\varphi_i(\hat{r}_i)R^T(\hat{y}_i - y_i) - k_{\theta i}\hat{\theta}_i], \tag{6.11}$$

where R, k_{θ_i} and $\gamma_{\theta i}$ are positive constants.

Define $\varpi_i = \hat{r}_i - r_i$, then

$$\dot{\varpi}_i =(A - LC)\varpi_i + B(\varphi_i^T(\hat{r}_i)\hat{\theta}_i - \varphi_i^T(r_i)\theta_i - o_i). \tag{6.12}$$

It shows that

$$\dot{\varpi} =(I_N \otimes (A - LC))\varpi + (I_N \otimes B)(\varphi^T(\hat{r})\hat{\theta} - \varphi^T(r)\theta - o), \tag{6.13}$$

where $\varpi = \text{col}\{\varpi_1, \varpi_2, \ldots, \varpi_N\}$, $\varphi(\hat{r}) = \text{diag}\{\varphi_1(\hat{r}_1), \varphi_2(\hat{r}_2), \ldots, \varphi_N(\hat{r}_N)\}$, $\hat{\theta} = \text{col}\{\hat{\theta}_1, \hat{\theta}_2, \ldots, \hat{\theta}_N\}$, $\varphi(r) = \text{diag}\{\varphi_1(r_1), \varphi_2(r_2), \ldots, \varphi_N(r_N)\}$, $\theta = \text{col}\{\theta_1, \theta_2, \ldots, \theta_N\}$, $o = \text{col}\{o_1, o_2, \ldots, o_N\}$.

Define the following parameters:

$$\alpha_i = \sum_{j=1}^{N} a_{ij}(\hat{r}_i - \hat{r}_j) + \sum_{d=1}^{2} g_i(\hat{r}_i - r_m(t)), i = 1, \ldots, N. \tag{6.14}$$

$$\hat{\alpha}_i = \sum_{j=1}^{N} a_{ij}(\hat{\mathcal{R}}_i - \hat{\mathcal{R}}_j) + \sum_{d=1}^{2} g_i(\hat{\mathcal{R}}_i - r_m(t)), i = 1, \ldots, N. \tag{6.15}$$

where $\hat{\mathcal{R}}_i = e^{A(t-t_k^i)}\hat{r}_i(t_k^i)$, t_k^i is determined later.

The global vector $\hat{\alpha}$ is represented as

$$\alpha = (\mathfrak{L}_1 \otimes I_n)\hat{r} + (\mathfrak{L}_2 \otimes I_n)r_{ref}, \tag{6.16}$$

$$\hat{\alpha} = (\mathfrak{L}_1 \otimes I_n)\hat{\mathcal{R}} + (\mathfrak{L}_2 \otimes I_n)r_{ref}, \tag{6.17}$$

where $\hat{\alpha} = \text{col}\{\hat{\alpha}_1(t), \hat{\alpha}_2(t), \ldots, \hat{\alpha}_N(t)\}$, $\hat{\mathcal{R}} = \text{col}\{\hat{\mathcal{R}}_1(t), \hat{\mathcal{R}}_2(t), \ldots, \hat{\mathcal{R}}_N(t)\}$.

Moreover, define

$$\varsigma = \hat{r} + (\mathfrak{L}_1^{-1}\mathfrak{L}_2 \otimes I_n)r_{ref}, \tag{6.18}$$

$$\hat{\varsigma} = \hat{\mathcal{R}} + (\mathfrak{L}_1^{-1}\mathfrak{L}_2 \otimes I_n)r_{ref}, \tag{6.19}$$

where $\varsigma = \mathrm{col}\{\varsigma_1(t), \varsigma_2(t), \ldots, \varsigma_N(t)\}$, $\hat{\varsigma} = \mathrm{col}\{\hat{\varsigma}_1(t), \hat{\varsigma}_2(t), \ldots, \hat{\varsigma}_N(t)\}$, $\hat{r} = \mathrm{col}\{\hat{r}_1(t), \hat{r}_2(t), \ldots, \hat{r}_N(t)\}$.

Based on the observer triggering states of neighbouring agents, we propose a distributed containment protocol to each follower. Let the input vector u_i in (6.4) be

$$u_i = -c_i \rho_i(t_k^i) B^T P \hat{\alpha}_i, \tag{6.20}$$

where $c_i, i = 1, 2, \ldots, N$ are positive constants, and

$$\dot{\rho}_i = -k_{\rho_i}(\rho_i - 1) + \gamma_{\rho_i}\hat{\alpha}_i^T PBB^T P\hat{\alpha}_i, \tag{6.21}$$

where k_{ρ_i} and γ_{ρ_i} are positive constants, P is a positive definite matrix to be determined later, and $\rho_i(0) \geq 1$.

Then, it is easy to obtain

$$\dot{\varsigma} = (I_N \otimes A)\varsigma - [\mathcal{C}\check{\Pi} \otimes BB^T P]\hat{\alpha} + (I_N \otimes B)\varphi^T(\hat{r})\hat{\theta} - (I_N \otimes LC)\varpi, \tag{6.22}$$

where $\mathcal{C} = \mathrm{diag}\{c_1, c_2, \ldots, c_N\}$, $\check{\Pi} = \mathrm{diag}\{\rho_1(t_k^1), \rho_2(t_k^2), \ldots, \rho_N(t_k^N)\}$. Moreover,

$$\dot{\alpha} = (I_N \otimes A)\alpha - [\mathfrak{L}_1\mathcal{C}\check{\Pi} \otimes BB^T P]\hat{\alpha} - (\mathfrak{L}_1 \otimes LC)\varpi + (\mathfrak{L}_1 \otimes B)\varphi^T(\hat{r})\hat{\theta}. \tag{6.23}$$

Moreover, t_k^i denotes the kth triggering instant of ith DG, which is dictated by the following trigger conditions:

$$t_{k+1}^i = \inf\{t > t_k^i : o_i \leq 0\}. \tag{6.24}$$

Further, define a dynamic variable $\dot{o}_i$ governed by

$$\dot{o}_i = \begin{cases} \min\{\mathcal{Y}_i, 0\} - \sigma_i, if\ \|\breve{e}_i(t)\| \neq 0 \\ -\sigma_i, if\ \|\breve{e}_i(t)\| = 0 \end{cases} \tag{6.25}$$

for $t \in [t_k^i, t_{k+1}^i]$, with

$$\begin{aligned} \mathcal{Y}_i = & -\frac{1}{2\aleph}\Big\{\bar{\lambda}(\Gamma)\Big[(c_i + 2)\rho_i + \vartheta_i\Big] + (\bar{\lambda}(P)\hbar + 1)o_i + (\iota_i + c_i^2)\hat{\alpha}_i^T PBB^T P\hat{\alpha}_i \\ & + (4\bar{\lambda}(\Gamma) + \gamma_{\rho_i}\hat{\alpha}_i^T PBB^T P\hat{\alpha}_i + k_{\rho_i}\pi_5)o_i^2\Big\} + \frac{\chi_i}{2\aleph\breve{e}_i^T\breve{e}_i}\hat{\alpha}_i^T PBB^T P\hat{\alpha}_i \end{aligned} \tag{6.26}$$

and $o_i(t_k^{i+}) = \nu_i, \sigma_i > 0, \nu_i > 0$ are designed parameters. Moreover, $\dot{\vartheta}_i = -k_{\vartheta_i}\vartheta_i + \gamma_{\vartheta_i} e_i^T PBB^T Pe_i$, $e_i(t)$ is the combined measurement error denoted as $e_i(t) = \hat{\alpha}_i - \alpha_i$. $\dot{\iota}_i = -k_{\iota_i}\iota_i + \gamma_{\iota_i}\acute{\rho}_i^2\hat{\alpha}_i^T PBB^T P\hat{\alpha}_i$, where $\acute{\rho}_i = \rho_i(t_k^i) - \rho_i(t)$. Denote $\check{e}_i = \text{col}\{e_i, \acute{\rho}\}$, $\check{P} = \text{diag}\{P, 1\}$.

For $t = t_k^i$, the jump dynamics are $\varpi_i(t_k^i+) = \varpi_i(t_k^i-)$, $\tilde{\theta}_i(t_k^i+) = \tilde{\theta}_i(t_k^i-)$, $\varsigma_i(t_k^i+) = \varsigma_i(t_k^i-)$, $\rho_i(t_k^i+) = \rho_i(t_k^i-)$, $\vartheta_i(t_k^i+) = \vartheta_i(t_k^i-)$, $\iota_i(t_k^i+) = \iota_i(t_k^i-)$, $\check{e}_i(t_k^i+) = 0$, $o_i(t_k^i+) = \nu_i$.

Remark 6.3 It is worth noting that the mechanism (6.24) proposed in this chapter implements the triggering behavior as a timer. The essence of mechanism (6.24) is that the timer variable $o_i(t)$ decreases from the maximum value $\nu_i > 0$ to 0, and an event is triggered when $o_i(t) = 0$. This also means that the dynamics satisfies $0 \le o_i(t) \le \nu_i$ and $\dot{\nu}_i \le 0$ throughout the process.

Theorem 6.1 *Under Assumptions 6.1, it shows that there exist $P > 0, \Psi > 0, \mathcal{M} > 0$, G and L with appropriate dimensions such that*

$$A^T\Psi + \Psi A + (\pi_2 + \pi_3)\Psi BB^T\Psi - 2C^T C + \pi_1\mathcal{B}\mathcal{B}^T + (\bar{\lambda}(\mathfrak{L}_1)^2 + 1)I_n + \mathcal{M} < 0, \tag{6.27}$$

$$PA^T + AP - PBB^T P + I_n = 0, \tag{6.28}$$

where $\mathcal{B} = \mathcal{H}_* C^T R + \Psi B$, $L = \Psi^{-1}C^T$, $\mathcal{H}_*$ is positive constant which is determined later. The controller (6.20) uses the proposed event-triggering strategy, and the consensus error of the critical bus voltage value of sub-MGs in a multi-MG system will converge exponentially to a bounded domain under the triggering conditions (6.24).

Proof Consider a Lyapunov function

$$V = \sum_{i=1}^N \varpi_i^T \Psi \varpi_i + \sum_{i=1}^N \frac{\mathcal{H}_*\tilde{\theta}_i^T\tilde{\theta}_i}{2\gamma_{\theta_i}} + \varsigma^T(\mathfrak{L}_1 \otimes P)\varsigma + \sum_{i=1}^N \frac{(\rho_i - \bar{\rho})^2}{2\gamma_{\rho_i}} + \sum_{i=1}^N \frac{(\vartheta_i - \bar{\vartheta})^2}{2\gamma_{\vartheta_i}} + \sum_{i=1}^N \frac{(\iota_i - \bar{\iota})^2}{2\gamma_{\iota_i}} + \check{e}(o \otimes \check{P})\check{e}, \tag{6.29}$$

where $\tilde{\theta}_i = \hat{\theta}_i - \theta_i$, and $\mathcal{H}_*, \bar{\rho}, \bar{\vartheta}, \bar{\iota}$ are positive constants, which are designed later. Because of $\rho_i(0) \ge 1$ and $\dot{\rho}_i \ge 0$, we can obtain that $\rho_i(t) \ge 1$ for any $t > 0$. Evidently, V is positive definite.

The time derivative of V can be obtained as follows:

$$\dot{V} = 2\sum_{i=1}^{N} \varpi_i^T \Psi \dot{\varpi}_i + \sum_{i=1}^{N} \frac{\mathcal{H}_* \tilde{\theta}_i^T \dot{\hat{\theta}}_i}{\gamma_{\theta_i}} + 2\varsigma^T (\mathfrak{L}_1 \otimes P)\dot{\varsigma} + \sum_{i=1}^{N} \frac{(\rho_i - \bar{\rho})\dot{\rho}_i}{\gamma_{\rho_i}}$$
$$+ \sum_{i=1}^{N} \frac{(\vartheta_i - \bar{\vartheta})\dot{\vartheta}_i}{\gamma_{\vartheta_i}} + \sum_{i=1}^{N} \frac{(\iota_i - \bar{\iota})\dot{\iota}_i}{\gamma_{\iota_i}} + 2\check{e}(o \otimes \check{P})\dot{\check{e}}. \tag{6.30}$$

Define $\tilde{\varphi} = \varphi_i(\hat{r}_i) - \varphi_i(r_i)$. Using Young's inequality yields, It is straightforward to show that

$$2\sum_{i=1}^{N} \varpi_i^T \Psi \dot{\varpi}_i + \sum_{i=1}^{N} \frac{\mathcal{H}_* \tilde{\theta}_i^T \dot{\hat{\theta}}_i}{\gamma_{\theta_i}}$$
$$= \sum_{i=1}^{N} \varpi_i^T \Big[(A - LC)^T \Psi + \Psi(A - LC)\Big]\varpi_i + 2\sum_{i=1}^{N} \varpi_i^T \mathcal{B}\varphi_i^T(\hat{r}_i)\tilde{\theta}_i$$
$$+ 2\sum_{i=1}^{N} \varpi_i^T \Psi B(\tilde{\varphi}_i^T \theta_i - o_i) - \sum_{i=1}^{N} \mathcal{H}_* k_{\theta_i} \tilde{\theta}_i^T \hat{\theta}_i$$
$$\leq \sum_{i=1}^{N} \varpi_i^T \Big[(A - LC)^T \Psi + \Psi(A - LC)\Big]\varpi_i + \pi_1 \sum_{i=1}^{N} \varpi_i^T \mathcal{B}\mathcal{B}^T \varpi_i$$
$$+ \pi_1^{-1} \sum_{i=1}^{N} \tilde{\theta}_i^T \varphi_i(\hat{r}_i)\varphi_i^T(\hat{r}_i)\tilde{\theta}_i + (\pi_2 + \pi_3)\sum_{i=1}^{N} \varpi_i^T \Psi B B^T \Psi \varpi_i$$
$$+ \pi_2^{-1} \sum_{i=1}^{N} \theta_i^T \tilde{\varphi}\tilde{\varphi}_i^T \theta_i + \pi_3^{-1} \sum_{i=1}^{N} o_i^T o_i - \sum_{i=1}^{N} \mathcal{H}_* k_{\theta_i} \tilde{\theta}_i^T \tilde{\theta}_i + \sum_{i=1}^{N} \mathcal{H}_* k_{\theta_i} \theta_i^T \theta_i. \tag{6.31}$$

Because of Lemma 2, from (6.31) one can obtain that

$$2\sum_{i=1}^{N} \varpi_i^T \Psi \dot{\varpi}_i + \sum_{i=1}^{N} \frac{\mathcal{H}_* \tilde{\theta}_i^T \dot{\hat{\theta}}_i}{\gamma_{\theta_i}}$$
$$\leq \sum_{i=1}^{N} \varpi_i^T \Big[A^T \Psi + \Psi A - 2C^T C + \pi_1 \mathcal{B}\mathcal{B}^T + (\pi_2 + \pi_3)\Psi B B^T \Psi\Big]\varpi_i$$
$$+ \sum_{i=1}^{N} (\mathcal{H}_* k_{\theta_i} + 4\pi_2^{-1} l_i)\theta_i^T \theta_i + \pi_3^{-1} \sum_{i=1}^{N} o_i^2 - \sum_{i=1}^{N} (\mathcal{H}_* k_{\theta_i} - \pi_1^{-1} l_i)\tilde{\theta}_i^T \tilde{\theta}_i. \tag{6.32}$$

Consider the following equalities

$$
\begin{aligned}
&2\varsigma^T(\mathfrak{L}_1 \otimes P)\dot{\varsigma} + \sum_{i=1}^{N} \frac{(\rho_i - \bar{\rho})\dot{\rho}_i}{\gamma_{\rho_i}} + \sum_{i=1}^{N} \frac{(\vartheta_i - \bar{\vartheta})\dot{\vartheta}_i}{\gamma_{\vartheta_i}} \\
&+ \sum_{i=1}^{N} \frac{(\iota_i - \bar{\iota})\dot{\iota}_i}{\gamma_{\iota_i}} + 2\breve{e}(o \otimes \breve{P})\dot{\breve{e}} + 2\breve{e}(\dot{o} \otimes \breve{P})\breve{e} \\
&= \varsigma^T\Big[\mathfrak{L}_1 \otimes (PA + A^T P)\Big]\varsigma - 2\varsigma^T(\mathfrak{L}_1 C\breve{\Pi} \otimes PBB^T P)\hat{\alpha} \\
&- 2\varsigma^T(\mathfrak{L}_1 \otimes PLC)\varpi + 2\varsigma^T(\mathfrak{L}_1 \otimes PB)\varphi^T(\hat{r})\hat{\theta} \\
&+ \sum_{i=1}^{N}(\rho_i - \bar{\rho})\hat{\alpha}_i^T PBB^T P\hat{\alpha}_i - \sum_{i=1}^{N} \frac{k_{\rho_i}}{\gamma_{\rho_i}}(\rho_i - \bar{\rho})(\rho_i - 1) \\
&+ \sum_{i=1}^{N}(\vartheta_i - \bar{\vartheta})e_i^T PBB^T Pe_i - \sum_{i=1}^{N} \frac{k_{\vartheta_i}}{\gamma_{\vartheta_i}}(\vartheta_i - \bar{\vartheta})\vartheta_i \\
&+ \sum_{i=1}^{N}(\iota_i - \bar{\iota})\acute{\rho}_i^2\hat{\alpha}_i^T PBB^T P\hat{\alpha}_i - \sum_{i=1}^{N} \frac{k_{\iota_i}}{\gamma_{\iota_i}}(\iota_i - \bar{\iota})\iota_i \\
&+ 2e(o \otimes P)\dot{e} + 2e(\dot{o} \otimes P)e + 2\sum_{i=1}^{N} o_i\acute{\rho}_i^T\dot{\acute{\rho}}_i + 2\sum_{i=1}^{N} \dot{o}_i\acute{\rho}_i^T\acute{\rho}_i.
\end{aligned}
\tag{6.33}
$$

Define $\acute{\Pi} = \{\acute{\rho}_1, \acute{\rho}_2, \ldots, \acute{\rho}_N\}$, because of (6.16) and (6.18), we can obtain $\alpha = (\mathfrak{L}_1 \otimes I_2)\varsigma$, then

$$
\begin{aligned}
&- 2\varsigma^T(\mathfrak{L}_1 C\breve{\Pi} \otimes PBB^T P)\hat{\alpha} \\
&= -2\varsigma^T(\mathfrak{L}_1 C\Pi \otimes PBB^T P)\hat{\alpha} - 2\varsigma^T(\mathfrak{L}_1 C\acute{\Pi} \otimes PBB^T P)\hat{\alpha} \\
&\leq -\sum_{i=1}^{N} c_i\rho_i\alpha_i^T PBB^T P\alpha_i + \sum_{i=1}^{N} c_i\rho_i e_i^T PBB^T Pe_i \\
&\quad + \sum_{i=1}^{N} \alpha_i^T PBB^T P\alpha_i + \sum_{i=1}^{N} c_i^2\acute{\rho}_i^2\hat{\alpha}_i^T PBB^T P\hat{\alpha}_i.
\end{aligned}
\tag{6.34}
$$

Then, it is easy to obtain

$$
\begin{aligned}
&- 2\varsigma^T(\mathfrak{L}_1 \otimes PLC)\varpi \\
&\leq \sum_{i=1}^{N} \alpha_i^T(t)PLCC^T L^T P\alpha_i(t) + \sum_{i=1}^{N} \varpi_i^T\varpi_i,
\end{aligned}
\tag{6.35}
$$

and

$$
\begin{aligned}
&2\varsigma^T(\mathfrak{L}_1 \otimes PB)\varphi^T(\hat{r})\hat{\theta} \\
&= 2\sum_{i=1}^{N}\alpha_i^T PB\varphi_i^T(\hat{r}_i)\tilde{\theta}_i + 2\sum_{i=1}^{N}\alpha_i^T PB\varphi_i^T(\hat{r}_i)\theta_i \\
&\leq 2\sum_{i=1}^{N}\alpha_i^T PBB^T P\alpha_i + \sum_{i=1}^{N} l_i\tilde{\theta}_i^T\tilde{\theta}_i + \sum_{i=1}^{N} l_i\theta_i^T\theta_i.
\end{aligned}
\tag{6.36}
$$

Then, we have the following inequalities

$$
\begin{aligned}
&\sum_{i=1}^{N}(\rho_i - \bar{\rho})\hat{\alpha}_i^T PBB^T P\hat{\alpha}_i \\
&\leq 2\sum_{i=1}^{N}\rho_i\alpha_i^T PBB^T P\alpha_i + 2\sum_{i=1}^{N}\rho_i e_i^T PBB^T Pe_i \\
&\quad - \frac{(1-\pi_4)}{2}\sum_{i=1}^{N}\bar{\rho}\alpha_i^T PBB^T P\alpha_i \\
&\quad + \frac{1}{2\pi_4}\sum_{i=1}^{N}\bar{\rho}\|B^T Pe_i(t)\|^2 - \sum_{i=1}^{N}\frac{\bar{\rho}}{2}\hat{\alpha}_i^T PBB^T P\hat{\alpha}_i.
\end{aligned}
\tag{6.37}
$$

Because of

$$
\dot{e} = -(I_N \otimes A)e + \left[\mathcal{C}\mathfrak{L}_1\acute{\Pi} \otimes BB^T P\right]\hat{\alpha} + (\mathfrak{L}_1 \otimes LC)\varpi - (\mathfrak{L}_1 \otimes B)\varphi^T(\hat{r})\hat{\theta},
\tag{6.38}
$$

we have

$$
\begin{aligned}
&2e(o \otimes P)\dot{e} \\
&= \sum_{i=1}^{N} o_i e_i^T(PA + A^T P)e_i + 2e(o\mathcal{C}\mathfrak{L}_1\acute{\Pi} \otimes PBB^T P)\hat{\alpha} \\
&\quad + 2e(o\mathfrak{L}_1 \otimes PLC)\varpi - 2e(o\mathfrak{L}_1 \otimes PB)\varphi^T(\hat{r})\hat{\theta}.
\end{aligned}
\tag{6.39}
$$

Then, consider the following equalities

$$
\begin{aligned}
&2e(o\mathcal{C}\mathfrak{L}_1\acute{\Pi} \otimes PBB^T P)\hat{\alpha} \\
&\leq \sum_{i=1}^{N} c_i^2 o_i^2 e_i^T PBB^T Pe_i + \bar{\lambda}(\mathfrak{L}_1)^2\sum_{i=1}^{N}\acute{\rho}_i^2\hat{\alpha}_i^T PBB^T P\hat{\alpha}_i,
\end{aligned}
\tag{6.40}
$$

$$
\begin{aligned}
&2e(o\mathfrak{L}_1 \otimes PLC)\varpi \\
&\quad \leq \sum_{i=1}^{N} o_i^2 e_i^T PLCC^T L^T Pe_i + \bar{\lambda}(\mathfrak{L}_1)^2 \sum_{i=1}^{N} \varpi_i^T \varpi_i.
\end{aligned} \tag{6.41}
$$

and

$$
\begin{aligned}
&-2e(o\mathfrak{L}_1 \otimes PB)\varphi^T(\hat{r})\hat{\theta} \\
&\quad \leq 2\sum_{i=1}^{N} o_i^2 e_i^T PBB^T Pe_i + \bar{\lambda}(\mathfrak{L}_1)^2 \sum_{i=1}^{N} l_i \tilde{\theta}_i^T \tilde{\theta}_i + \bar{\lambda}(\mathfrak{L}_1)^2 \sum_{i=1}^{N} l_i \theta_i^T \theta_i.
\end{aligned} \tag{6.42}
$$

By substituting (6.40)–(6.42) into (6.39), we have

$$
\begin{aligned}
&2e(o \otimes P)\dot{e} \\
&\leq \sum_{i=1}^{N} o_i e_i^T e_i + 4\sum_{i=1}^{N} o_i^2 e_i^T PBB^T Pe_i + \bar{\lambda}(\mathfrak{L}_1)^2 \sum_{i=1}^{N} \varpi_i^T \varpi_i \\
&\quad + \bar{\lambda}(\mathfrak{L}_1)^2 \sum_{i=1}^{N} \rho_i^2 \hat{\alpha}_i^T PBB^T P\hat{\alpha}_i + \bar{\lambda}(\mathfrak{L}_1)^2 \sum_{i=1}^{N} l_i \tilde{\theta}_i^T \tilde{\theta}_i + \bar{\lambda}(\mathfrak{L}_1)^2 \sum_{i=1}^{N} l_i \theta_i^T \theta_i.
\end{aligned} \tag{6.43}
$$

Moreover, because of

$$
\dot{\rho}_i = k_{\rho_i}(\rho_i - 1) - \gamma_{\rho_i} \hat{\alpha}_i^T PBB^T P\hat{\alpha}_i, \tag{6.44}
$$

then, we have

$$
\begin{aligned}
&2\sum_{i=1}^{N} o_i \acute{\rho}_i^T \dot{\rho}_i \\
&\quad \leq 2\sum_{i=1}^{N} k_{\rho_i} o_i \acute{\rho}_i^T (\rho_i - 1) - 2\sum_{i=1}^{N} \gamma_{\rho_i} o_i \acute{\rho}_i^T \hat{\alpha}_i^T PBB^T P\hat{\alpha}_i \\
&\quad \leq \sum_{i=1}^{N} k_{\rho_i} \pi_5 o_i^2 \acute{\rho}_i^2 + \sum_{i=1}^{N} k_{\rho_i} \pi_5^{-1} (\rho_i - 1)^2 \\
&\qquad + \sum_{i=1}^{N} \hat{\alpha}_i^T PBB^T P\hat{\alpha}_i + \sum_{i=1}^{N} \gamma_{\rho_i} o_i^2 \acute{\rho}_i^2 \hat{\alpha}_i^T PBB^T P\hat{\alpha}_i.
\end{aligned} \tag{6.45}
$$

Because of $\bar{\iota} \geq \bar{\lambda}(\mathfrak{L}_1)^2 \sum_{i=1}^{N} c_i^2$, $\bar{\vartheta} \geq \frac{1}{2\pi_4}\bar{\rho}$, $\bar{\rho} \geq 2\chi_i + 2$, define $\aleph = \min\{1, \underline{\lambda}(P)\}$, $\Gamma = PBB^TP$, by substituting (6.34)–(6.37) and (6.43)–(6.45) into (6.33), one can deduce

$$\begin{aligned}
&2\varsigma^T(\mathfrak{L}_1 \otimes P)\dot{\varsigma} + \sum_{i=1}^{N} \frac{(\rho_i - \bar{\rho})\dot{\rho}_i}{\gamma_{\rho_i}} + \sum_{i=1}^{N} \frac{(\vartheta_i - \bar{\vartheta})\dot{\vartheta}_i}{\gamma_{\vartheta_i}} + \sum_{i=1}^{N} \frac{(\iota_i - \bar{\iota})\dot{\iota}_i}{\mathcal{M}_* \gamma_{\iota_i}} + \check{e}(o \otimes \check{P})\dot{\check{e}} \\
&\leq \varsigma^T\Big[\mathfrak{L}_1 \otimes (PA + A^TP)\Big]\varsigma \\
&\quad - \sum_{i=1}^{N}\Big[(c_i - 2)\rho_i - 4 + \frac{(1-\pi_4)}{2}\bar{\rho}\Big]\alpha_i^T(t)PBB^TP\alpha_i(t) \\
&\quad + \sum_{i=1}^{N}\Big\{\bar{\lambda}(\Gamma)\Big[(c_i + 2)\rho_i + \vartheta_i + 4o_i^2\Big] + o_i\Big\}e_i^T(t)e_i(t) \\
&\quad + \sum_{i=1}^{N}\Big[(\gamma_{\rho_i} o_i^2 + \iota_i + c_i^2)\hat{\alpha}_i^T PBB^TP\hat{\alpha}_i + k_{\rho_i}\pi_5 o_i^2\Big]\acute{\rho}_i^2 \\
&\quad - \sum_{i=1}^{N}\chi_i\hat{\alpha}_i^T PBB^TP\hat{\alpha}_i + 2\aleph\sum_{i=1}^{N}\dot{o}_i\check{e}_i^T\check{e}_i + \hbar\sum_{i=1}^{N}o_i\check{e}_i^T P\check{e}_i - \hbar\sum_{i=1}^{N}o_i\check{e}_i^T P\check{e}_i \\
&\quad - \sum_{i=1}^{N}\frac{k_{\rho_i}}{\gamma_{\rho_i}}(\rho_i - \bar{\rho})(\rho_i - 1) - \sum_{i=1}^{N}\frac{k_{\vartheta_i}}{\gamma_{\vartheta_i}}(\vartheta_i - \bar{\vartheta})\vartheta_i - \sum_{i=1}^{N}\frac{k_{\iota_i}}{\gamma_{\iota_i}}(\iota_i - \bar{\iota})\iota_i \\
&\quad + \sum_{i=1}^{N}k_{\rho_i}\pi_5^{-1}(\rho_i - 1)^2 + (\bar{\lambda}(\mathfrak{L}_1)^2 + 1)\sum_{i=1}^{N}\Big[l_i(\tilde{\theta}_i^T\tilde{\theta}_i + \theta_i^T\theta_i) + \varpi_i^T\varpi_i\Big].
\end{aligned} \tag{6.46}$$

Because of $c_i \geq 2$ and $\bar{\rho} \geq 2(\lambda_{\max}(\mathfrak{L}_1) + 6 - c_i)/(1 - \pi_4)$, such that

$$\begin{aligned}
\dot{V} \leq &\sum_{i=1}^{N}\varpi_i^T\Big[A^T\Psi + \Psi A - 2C^TC + \pi_1\mathcal{B}\mathcal{B}^T + (\pi_2 + \pi_3)\Psi BB^T\Psi \\
&+ (\bar{\lambda}(\mathfrak{L}_1)^2 + 1)\Big]\varpi_i - \varsigma^T(\mathfrak{L}_1 \otimes I_N)\varsigma \\
&+ \sum_{i=1}^{N}\Big\{\bar{\lambda}(\Gamma)\Big[(c_i + 2)\rho_i + \vartheta_i\Big] + (\bar{\lambda}(P)\hbar + 1)o_i + (\iota_i + c_i^2)\hat{\alpha}_i^T PBB^TP\hat{\alpha}_i \\
&+ (4\bar{\lambda}(\varGamma) + \gamma_{\rho_i}\hat{\alpha}_i^T PBB^TP\hat{\alpha}_i + k_{\rho_i}\pi_5)o_i^2\Big\}\check{e}_i^T\check{e}_i \\
&- \sum_{i=1}^{N}\chi_i\hat{\alpha}_i^T PBB^TP\hat{\alpha}_i + 2\aleph\sum_{i=1}^{N}\dot{o}_i\check{e}_i^T\check{e}_i - \hbar\sum_{i=1}^{N}o_i\check{e}_i^T P\check{e}_i
\end{aligned}$$

$$
\begin{aligned}
&+\sum_{i=1}^{N}\Big[\mathcal{H}_* k_{\theta_i}+4\pi_2^{-1}l_i+(\bar{\lambda}(\mathfrak{L}_1)^2+1)l_i\Big]\theta_i^T\theta_i\\
&+\pi_3^{-1}\sum_{i=1}^{N}o_i^2-\sum_{i=1}^{N}\Big[\mathcal{H}_* k_{\theta_i}-(\bar{\lambda}(\mathfrak{L}_1)^2+1+\pi_1^{-1})l_i\Big]\tilde{\theta}_i^T\tilde{\theta}_i\\
&-\sum_{i=1}^{N}\frac{k_{\rho_i}}{\gamma_{\rho_i}}(\rho_i-\bar{\rho})(\rho_i-1)+\sum_{i=1}^{N}k_{\rho_i}\pi_5^{-1}(\rho_i-1)^2\\
&-\sum_{i=1}^{N}\frac{k_{\vartheta_i}}{\gamma_{\vartheta_i}}(\vartheta_i-\bar{\vartheta})\vartheta_i-\sum_{i=1}^{N}\frac{k_{\iota_i}}{\gamma_{\iota_i}}(\iota_i-\bar{\iota})\iota_i. \qquad (6.47)
\end{aligned}
$$

According to the triggering condition (6.24), $\mathcal{H}_* \geq k_{\theta_i}^{-1}\Big[z_i+(\bar{\lambda}(\mathfrak{L}_1)^2+1+\pi_1^{-1})l_i\Big]$, $\pi_5 > 2\gamma_{\rho_i}$, we obtain

$$
\begin{aligned}
\dot{V}\leq &-\delta V-\varsigma^T\Big[\mathfrak{L}_1\otimes(I_N-\delta P)\Big]\varsigma-\varpi^T(\mathcal{M}-\delta\Psi)\varpi\\
&+\sum_{i=1}^{N}\Big[\mathcal{H}_* k_{\theta_i}+4\pi_2^{-1}l_i+(\bar{\lambda}(\mathfrak{L}_1)^2+1)l_i\Big]\theta_i^T\theta_i+\pi_3^{-1}\sum_{i=1}^{N}o_i^2\\
&-\sum_{i=1}^{N}\Big[k_{\rho_i}(1-2\gamma_{\rho_i}\pi_5^{-1})-\delta\Big]\frac{(\rho_i(t)-\bar{\rho})^2}{2\gamma_{\rho_i}}-\sum_{i=1}^{N}(k_{\vartheta_i}-\delta)\frac{(\vartheta_i(t)-\bar{\vartheta})^2}{2\gamma_{\vartheta_i}}\\
&+\sum_{i=1}^{N}\frac{k_{\vartheta_i}}{2\gamma_{\vartheta_i}}\bar{\vartheta}^2-\sum_{i=1}^{N}(z_i-\frac{\delta\mathcal{H}_*}{2\gamma_{\theta_i}})\tilde{\theta}_i^T\tilde{\theta}_i+(\delta-\hbar)\check{e}(o\otimes\check{P})\check{e}\\
&-\sum_{i=1}^{N}(k_{\iota_i}-\delta)\frac{(\iota_i(t)-\bar{\iota})^2}{2\gamma_{\iota_i}}+\sum_{i=1}^{N}\frac{k_{\iota_i}}{2\gamma_{\iota_i}}\bar{\iota}^2+2\sum_{i=1}^{N}k_{\rho_i}(\pi_5^{-1}+\frac{1}{4\gamma_{\rho_i}})(\bar{\rho}-1)^2. \qquad (6.48)
\end{aligned}
$$

Define $0<\delta\leq\min\{\frac{1}{\lambda_{\max}(P)},\frac{1}{\lambda_{\max}(M)},k_{\rho_i}(1-2\gamma_{\rho_i}\pi_5^{-1}),k_{\vartheta_i},k_{\iota_i},\frac{2\gamma_{\theta i}z_i}{\mathcal{H}_*},\hbar\}$, we obtain

$$
\dot{V}\leq -\delta V+\mathcal{Z}. \qquad (6.49)
$$

V exponentially converges to a bounded set

$$
\mathcal{D}_1 \triangleq \left\{\varpi_i,\tilde{\theta}_i,\rho_i,\vartheta_i,\iota_i,\varsigma,\check{e}: V\leq\frac{1}{\delta}\mathcal{Z}\right\}, \qquad (6.50)
$$

where

$$\mathcal{Z} = \sum_{i=1}^{N}[\mathcal{H}_* k_{\theta_i} + 4\pi_2^{-1} l_i + (\bar{\lambda}(\mathfrak{L}_1)^2 + 1) l_i \Big] \theta_i^T \theta_i + \pi_3^{-1} \sum_{i=1}^{N} o_i^2 + \sum_{i=1}^{N} \frac{k_{\vartheta_i}}{2\gamma_{\vartheta_i}} \bar{\vartheta}^2$$
$$+ \sum_{i=1}^{N} \frac{k_{\iota_i}}{2\gamma_{\iota_i}} \bar{\iota}^2 + 2\sum_{i=1}^{N} k_{\rho_i} (\pi_5^{-1} + \frac{1}{4\gamma_{\rho_i}})(\bar{\rho} - 1)^2. \tag{6.51}$$

Therefore $\varpi_i, \tilde{\theta}_i, \rho_i, \vartheta_i, \iota_i, i = 1, 2, \ldots, N$ and $\varsigma, \check{e}$ are uniformly ultimately bounded and exponentially converges to the residual set $\mathcal{D}_1$.

■

Remark 6.4 Due to the introduction of adaptive dynamics $\rho_i(t)$, ϑ_i, ι_i in the control protocol, the protocol no longer needs to include information related to parameters such as the communication network structure and the number of DGs. The protocol is therefore implemented in a fully distributed framework. This helps to reduce the bandwidth requirements and costs of the underlying communication network and enhance the flexibility and reliability of the overall MG operation.

Remark 6.5 The voltage regulation problem considered in this chapter is a containment-based control problem. The transformed linear system of the MG system contains nonlinear dynamics and the unpredictability of state $\dot{v}_{bi}$. The three problems mentioned above make it more challenging to design a trigger mechanism for the existence of a minimum event interval than [1]. Moreover, unlike [1], an innovative event-triggered mechanism relying on multiple measurement errors is designed to implement an inclusion control-based voltage regulation strategy, where the control protocol is updated intermittently.

Theorem 6.2 *The secondary voltage regulation strategy under Theorem 6.1 does not exhibit Zeno behavior for all agents at any finite time.*

Proof For DG i, the event interval $t_{k+1}^i - t_k^i$ is the period when o_i decreases from γ_i to 0. To calculate the uniform MIET of DG γ_i to 0, obtaining a lower bound on $\dot{o}_i$ is necessary. Consider the evolution of o_i over the time interval $[t_k^i, t_{k+1}^i)$. In the case when $\|e_i(t)\| \neq 0$,

$$\mathcal{Y}_i \geq -\frac{1}{2\aleph}\Big\{\bar{\lambda}(\Gamma)\Big[(c_i + 2)\rho_i + \vartheta_i\Big] + (\bar{\lambda}(P)\hbar + 1)o_i + (\iota_i + c_i^2)\hat{\alpha}_i^T PBB^T P\hat{\alpha}_i$$
$$+ \Big(4\bar{\lambda}(\Gamma) + \gamma_{\rho_i}\hat{\alpha}_i^T PBB^T P\hat{\alpha}_i + k_{\rho_i}\pi_5\Big)o_i^2\Big\}. \tag{6.52}$$

Because o_i is bounded over $[0, t_\infty)$, one has o_i

$$\mathcal{Y}_i \geq -\ell_i - \jmath_i o_i^2, \tag{6.53}$$

where $\ell_i = \frac{1}{2\aleph}\Big\{\bar{\lambda}(\Gamma)\Big[(c_i+2)\rho_i + \vartheta_i\Big] + \Big(\bar{\lambda}(P)\hbar + 1\Big)o_i + (\iota_i + c_i^2)\hat{\alpha}_i^T PBB^T P\hat{\alpha}_i\Big\}$, $\jmath_i = \frac{1}{2\aleph}\Big(4\bar{\lambda}(\Gamma) + \gamma_{\rho_i}\hat{\alpha}_i^T PBB^T P\hat{\alpha}_i + k_{\rho_i}\pi_5\Big)$. Thus, by (7), $\dot{o}_i$ is lower bounded by

$$\dot{o}_i \geq -\ell_i - \jmath_i\phi_i^2 - \sigma_i. \tag{6.54}$$

Therefore, by the comparison lemma, it is obtained that $o_i \geq \phi_i$, for $t \in [t_k^i, t_{k+1}^i)$, where ϕ_i satisfies the follow equation:

$$\dot{\phi}_i = -\ell_i - \jmath_i o_i^2 - \sigma_i, \tag{6.55}$$

with $\phi_i(t_k^{i+}) = \nu_i$. Solving equation (6.55), we obtain

$$t - t_k^i \geq \sqrt{\frac{1}{\jmath_i(\ell_i+\sigma_i)}}\Big[\arctan\Big(\sqrt{\frac{\jmath_i}{\ell_i+\sigma_i}}\nu_i\Big) - \arctan\Big(\sqrt{\frac{\jmath_i}{\ell_i+\sigma_i}}\phi_i\Big)\Big]. \tag{6.56}$$

In view of $o_i \geq \phi_i$, from (6.56), we obtain

$$\tau_i = t_{k+1}^i - t_k^i \geq \sqrt{\frac{1}{\jmath_i(\ell_i+\sigma_i)}}\arctan\Big(\sqrt{\frac{\jmath_i}{\ell_i+\sigma_i}}\nu_i\Big) > 0. \tag{6.57}$$

This implies that the Zeno phenomenon is excluded, that is, $\lim_{k\to\infty} t_k^i = \infty$, and hence $t_\infty = \infty$. Thus, a uniform lower bound of the inter-event times $t_{k+1}^i - t_k^i$ is given by $\tau_i > 0$, for any $i = 1, \ldots, N$.

Therefore, the above voltage regulation strategy can regulate the critical bus voltage to a reasonable setting range. And relying on the proposed trigger mechanism with positive MIET, the controller does not have to be updated continuously and saves computational costs.

6.4 Simulation Results

This section aims to demonstrate the feasibility and effectiveness of the proposed event-triggered secondary control strategy based on containment and output feedback for critical bus voltage co-regulation. For this purpose, we have simulated several scenarios in the MATLAB/Simulink software environment for the MG shown in Fig. 6.2. The multi-MG system is tested with a combination of four sub-MGs, where $v_{b1}, v_{b2}, v_{b3}, v_{b4}$ are the critical bus voltage of MG#1, MG#2, MG#3, MG#4, respectively. Table 6.1 provides the parameters of DGs, lines and loads in Table 6.1. It is assumed that the DGs communicate with each other through the directed topology shown in Fig. 6.2. DG#1, DG#4 receives information from v_{ref1}, v_{ref2}, respectively.

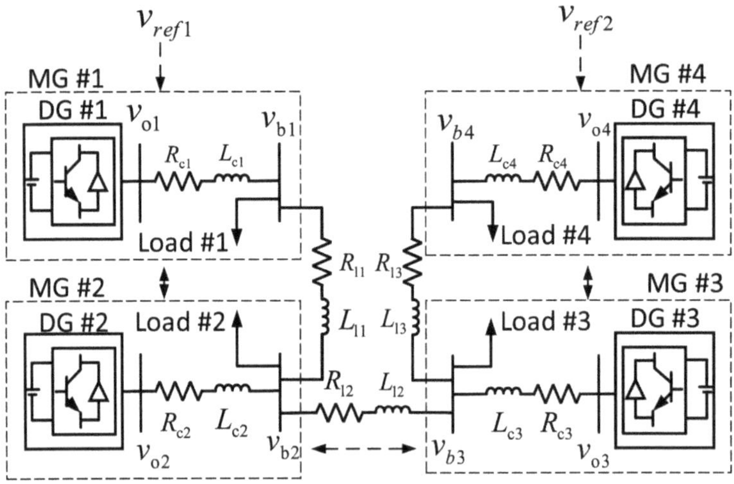

Fig. 6.2 The example MGs test system and considered communication structure

Table 6.1 Specifications of the MG test system

DGs	DG#1 & DG#2 & DG#3 & DG#4	
	m_P	1.5×10^{-5}
	n_Q	2×10^{-4}
Lines	$R_{l1} = R_{l3} = 1e - 4\Omega$	$R_{l2} = 1e - 4\Omega$
	$L_{l1} = L_{l3} = 0.318\,\text{emH}$	$L_{l2} = 1.847\,\text{mH}$
RL loads	Load #1 & Load #2& Load #3& Load #4	
	P(per phase)	Q(per phase)
	50 kW	120 kVAr

Solving (6.28) gives a solution

$$P = \begin{bmatrix} 1.7321 & 1.0000 \\ 1.0000 & 1.7321 \end{bmatrix}.$$

We define $v_{ref1} = 230$, $v_{ref2} = 210$, then, we obtain $\bar{r} = \text{col}\{226, 222, 218, 214\}$. Let $\rho_i(0.4) = 1$ and $\hat{\theta}_i(0.4) = 1$ be the initial value of $\rho_i(t)$ and $\hat{\theta}_i(t)$, $i = 1, 2, \ldots, N$, respectively. Moreover, $c_i = 1.5e + 7$, $k_{\rho_i} = k_{\vartheta_i} = k_{\iota_i} = 10$, $\gamma_{\rho_i} = \gamma_{\vartheta_i} = \gamma_{\iota_i} = 1$, $\sigma_i = 0.5$, $\nu_i = 40$, $i = 1, 2, 3, 4$.

To verify the effectiveness of the containment-based critical bus voltage algorithm, as shown in Fig. 6.3. We rely only on the primary control at the beginning. At t = 0.4 s, the secondary control of the containment-based critical bus voltage regulation cuts in, and the critical bus voltage returns to the convex hull spanned by all reference values. At $t = 0.75$ s, load #3 is cut off, and at t = 1.1 s, load #3 returns to its original value. There is some fluctuation in the load voltage, but after a short time, the load

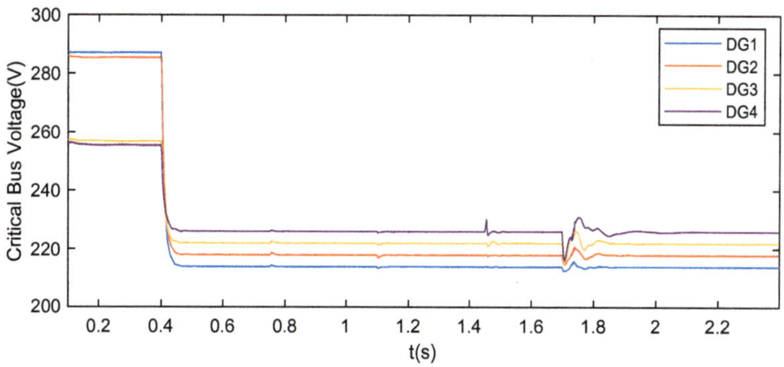

Fig. 6.3 The critical bus voltage

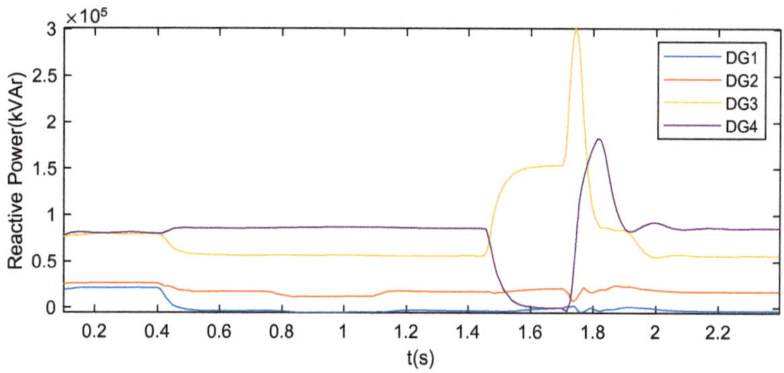

Fig. 6.4 Reactive power

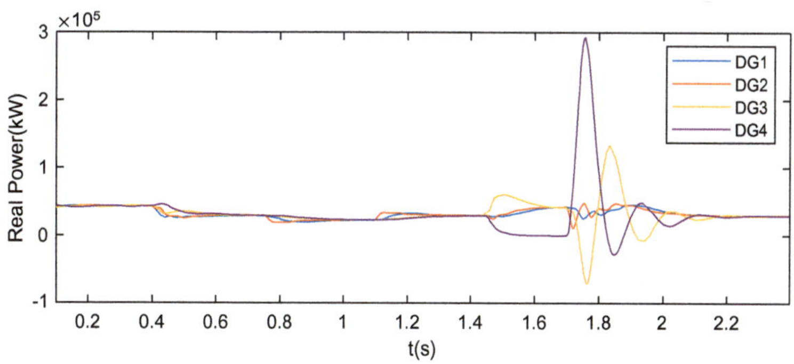

Fig. 6.5 Active power

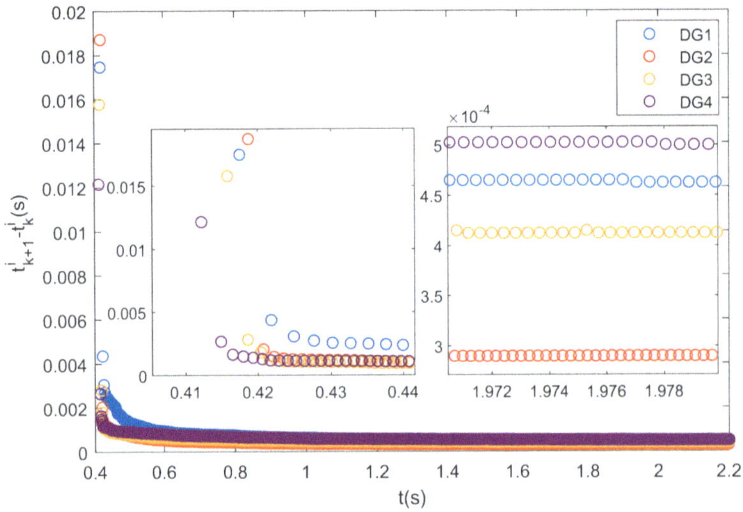

Fig. 6.6 Inter-event times $t^i_{k+1} - t^i_k, i = 1, 2, 3, 4$

voltage returns to the convex hull spanned by all reference values. The MG #4 is cut off at $t = 1.45$ s, and at t = 1.7 s, the MG #4 is restored. The voltage can still be restored.

The results of this simulation are shown in Figs. 6.3 and 6.4. In Fig. 6.5, the proposed control method can regulate critical bus voltage, the active power and reactive power after large-signal disturbances (i.e., load changes and a multi-MG system structure reconfiguration).

For clear visibility, the trigger time for each trigger of DG#1, DG#2, DG#3, DG#4 is shown in Fig. 6.6.

Therefore, the containment-based critical bus voltage regulation strategy proposed in this chapter proves effective, which also applies to the case of load switching and plug in & plug out capability.

6.5 Conclusions

In this chapter, a containment-based distributed critical bus voltage control strategy of a multi-MG system was proposed. Firstly, the critical bus voltage regulation problem is transformed into a linear MAS distributed output feedback tracking problem containing nonlinear dynamics. A fuzzy logic system approximates the unknown nonlinear dynamics. Next, a containment-based secondary controller was designed. It allows the critical bus voltages to be limited to a defined range while enabling the existence of voltage differences between sub-MGs, thus allowing power flow among them. In addition, a novel event-triggering mechanism was proposed. It is not only demonstrated that a uniform and strictly positive MIET exists for each DG under the triggering mechanism but also that successive updates of the secondary controller are avoided. Furthermore, this voltage control strategy does not require using any global information in the communication graph. Finally, the simulation verifies the results.

Reference

1. Qian, Y.Y., Wan, Y.: Design of distributed adaptive event-triggered consensus control strategies with positive minimum inter-event times. Automatica (2021)

MIX
Papier aus verantwortungsvollen Quellen
Paper from responsible sources
FSC® C105338

If you have any concerns about our products,
you can contact us on
ProductSafety@springernature.com

In case Publisher is established outside the EU,
the EU authorized representative is:
Springer Nature Customer Service Center GmbH
Europaplatz 3, 69115 Heidelberg, Germany

Printed by Libri Plureos GmbH
in Hamburg, Germany